Basic Concept of Analysis II with Conic Section

Basic Concept of Analysis II with Conic Section

S.P.Gupta

CYBER TECH PUBLICATIONS
4264/3, Ansari Road, Daryaganj, New Delhi-110002 (India)
Ph.: 011-23244078, 011-43559448 Fax: 011-23280028
E-Mail: cyberpublicationsdelhi@yahoo.com
Website: www.cybertechpublications.com

BASIC CONCEPT OF ANALYSIS II WITH CONIC SECTION

S.P.Gupta

First Edition 2010

Asian Offset Printers

Published by :

G.S. Rawat for **Cyber Tech Publications**
4264/3, Ansari Road, Daryaganj, New Delhi-110002 (India)
Ph.: 011-23244078, 011-43559448 Fax: 011-23280028
E-Mail: cyberpublicationsdelhi@yahoo.com
Website: www.cybertechpublications.com

PREFACE

The "Basic Concept of Analysis II with Conic Section" books has been specially written to the requirements syllabus of the students of all Indian Universities.

Each chapter of this book contains complete theory and fairly large number of solved examples. Sufficient problems has also been selected from various universities examination papers. and the author shall be thankful to the readers who point out errors and omissions which inspite of all care might have been there. The Subject matter has been discussed in such a simple way that the students will find no difficulty to understand it. The proofs of various theorems and examples has been given with minute details.

The author will feel amply rewarded if the book serve the purpose for which it is meant. Suggestions for the improvement of the book are always welcome.

—Author

Contents

1

Some Points of a Conic Section

INTRODUCTON

To find the equation of the chord of the conic $l/r = 1 + e \cos \theta$, *whose extremities are* (r_1, θ_1) *and* (r_2, θ_2).

$l/r = 1 + e \cos \theta$.

The equation of the conic is $l/r = 1 + e \cos \theta$. ...(1)

Let the points P (r_1, θ_1) and Q (r_2, θ_2) lie on (1). Then we have

$$l/r_1 = 1 + e \cos \theta_1 \quad \text{and} \quad l/r_2 = 1 + e \cos \theta_2.$$

Let the polar equation of any line be given as

$$A \cos \theta + B \sin \theta = l/r. \qquad \text{...(2)}$$

If the straight line (2) passes through the point P (r_1, θ_1), we have

$$A \cos \theta_1 + B \sin \theta_1 = l/r_1 = 1 + e \cos \theta_1$$

$$\Rightarrow \quad (A - e) \cos \theta_1 + B \sin \theta_1 - 1 = 0. \qquad \text{...(3)}$$

Similarly if the straight line (2) passes through the point Q (r_2, θ_2), we have

$$(A - e) \cos \theta_2 + B \sin \theta_2 - 1 = 0. \qquad \text{...(4)}$$

Solving (3) and (4) for A – e and B, we get

$$\frac{A-e}{\sin\theta_2 - \sin\theta_1} = \frac{B}{\cos\theta_1 - \cos\theta_2} = \frac{1}{\sin\theta_2 \sin\theta_1 - \cos\theta_2 \sin\theta_1}$$

$$= 1/\sin(\theta_2 - \theta_1)$$

$$\Rightarrow \quad \frac{A-e}{2\cos\frac{1}{2}(\theta_1+\theta_2)\sin\frac{1}{2}(\theta_2-\theta_1)} = \frac{A-e}{2\sin\frac{1}{2}(\theta_1+\theta_2)\sin\frac{1}{2}(\theta_2-\theta_1)}$$

$$= \frac{1}{2\sin\frac{1}{2}(\theta_2-\theta_1)\cos\frac{1}{2}(\theta_2-\theta_1)}$$

$$\Rightarrow \quad \frac{A-e}{\cos\frac{1}{2}(\theta_1+\theta_2)} = \frac{B}{\sin\frac{1}{2}(\theta_1+\theta_2)} = \frac{1}{\cos\frac{1}{2}(\theta_2-\theta_1)}$$

$$\therefore \quad A = \cos\frac{1}{2}(\theta_1+\theta_2)\sec\frac{1}{2}(\theta_2-\theta_1) + e,$$

$$B = \sin\frac{1}{2}(\theta_1+\theta_2)\sec\frac{1}{2}(\theta_2-\theta_1).$$

Putting these values of A and B in (2), the equation of the chord PQ is given by

$$l/r = \{\cos\frac{1}{2}(\theta_1+\theta_2)\sec\frac{1}{2}(\theta_2-\theta_1)+e\}\cos\theta + \{\sin\frac{1}{2}(\theta_1+\theta_2)\sec\frac{1}{2}(\theta_2-\theta_1)\}\sin\theta$$

$$\Rightarrow \quad 1/r = e\cos\theta + \sec\frac{1}{2}(\theta_2-\theta_1)\cos\{\theta-\frac{1}{2}(\theta_1+\theta_2)\} \qquad \text{...(5)}$$

Thus remember that (5) is the equation of the chord joining the points 'θ_1' and 'θ_2' on the conic $l/r = 1 + e\cos\theta$.

Cor. If the points P and Q are such that their vectorial angles are $(\alpha-\beta)$ and $\alpha+\beta$, so that the sum of the angles is 2α and their difference is 2β, then the equation (5) of the chord PQ becomes

$$1/r = e\cos\theta\sec\beta\cos(\theta-\alpha). \qquad \text{...(6)}$$

TANGENT TO THE CONIC AT A GIVEN POINT

To find the equation of the tangent at the point (r_1, θ_1) of the conic

$$l/r = 1 + e\cos\theta.$$

Let P be a given point (r_1, θ_1) on the conic $l/r = 1 + e\cos\theta$. Take another point Q (r_2, θ_2) on the conic. We get the equation of the chord joining the points P and Q as

$$l/r = e\cos\theta + \sec\frac{1}{2}(\theta_2-\theta_1)\cos\{\theta-\frac{1}{2}(\theta_1+\theta_2)\}.$$

Now the tangent at P to the conic $l/r = 1 + e\cos\theta$ is the limiting position of the chord PQ as $Q \to P$ *i.e.*, as $\theta_2 \to \theta_1$. So taking the limit of the equation of the chord PQ as $\theta_2 \to \theta_1$, we get the equation of the tangent to the conic $l/r = 1 + e \cos \theta$ at the point whose vectorial angle is θ_1 as

$$l/r = e \cos \theta + \cos (\theta - \theta_1). \qquad ...(7)$$

Cor. 1. If the conic is $l/r = 1 \cos (\theta - \alpha)$, the tangent at the point 'θ_1' is given by

$$l/r = e \cos (\theta - \alpha) + \cos (\theta - \theta_1). \qquad ...(8)$$

Cor. 2. If the conic is $l/r = 1 - e \cos \theta$, the tangent at the point θ_1 is

$$l/r = e \cos (\pi + \theta) + \cos \{(\pi + \theta) - (\pi + \theta_1)\}$$

i.e., $$l/r = - e \cos \theta + \cos (\theta - \theta_1). \qquad ...(9)$$

Cor. 3. To find the slope of the tangent (7). The equation (7) may be written as

$$l = r \cos \theta \cos \theta_1 + r \sin \theta \sin \theta_1 + e \,.\, r \cos \theta$$

$$\Rightarrow \quad l = r \cos \theta_1 + y \sin \theta_1 + ex$$

$$\Rightarrow$$

$$y \sin \theta_1 = - x (+ \cos \theta_1) + l.$$

$\therefore$ the slope of the tangent (7)

$$= - \frac{e + \cos\theta_1}{\sin\theta_1} \qquad ...(10)$$

ASYMPTOTES

To find the equation of the asymptotes of the conic $l/r = 1 + e \cos \theta$.

The equation of the conic is given by

$$l/r = 1 + e \cos \theta. \qquad ...(1)$$

Suppose (r', α) is a point on (1), so that we have

$$l/r' = 1 + e \cos \alpha. \qquad ...(2)$$

The equation of the tangent to the conic (1) at the point (r', α) on it is given as

$$l/r = e \cos \theta + \cos (\theta - \alpha). \qquad ...(3)$$

We know that an asymptote is the limiting position of the tangent as the point of contact tends to infinity. Hence (3) will tend to an asymptote if r' tends to infinity. Now if r' tends to infinity, we have from (2), $0 = 1 + e \cos \alpha$.

$$\therefore \quad \cos \alpha = - 1/e \text{ and } \sin \alpha = \pm \sqrt{(1 - 1/e^2)} \qquad ...(4)$$

From (3), we have

$$l/r = e \cos \theta + \cos \theta \cos \alpha + \sin \theta \sin \alpha.$$

Substituting for cos α and sin α from (4) in it, we have

$$l/r = e \cos \theta + \cos \theta (-1/e) \pm \sin \theta \sqrt{(1 - 1/e^2)}$$

$$\Rightarrow \quad le/r = (e^2 - 1) \cos \theta \pm \sqrt{(e^2 - 1)} \sin \theta. \quad ...(5)$$

The equation (5) represents the equations of the two asymptotes of the conic (1). Clearly the asymptotes are real only when e > 1.e., the conic (1) is a hyperbola.

Example 1:

Find the condition that the line $l/r = A \cos \theta + B \sin \theta$ may be a tangent to the conic $l/r = 1 + e \cos \theta$.

Solution:

Suppose the line $l/r = A \cos \theta + B \sin \theta$ ...(1)

is a tangent to the conic $l/r = 1 + e \cos \theta$ ...(2)

at the point whose vectorial angle is α. The equation of the tangent to (2) at the point 'a' is given by

$$l/r = \cos (\theta - \alpha) + e \cos \theta$$

$$\Rightarrow \quad l/r = (e + \cos \theta) \cos \theta + \sin \theta \sin \alpha. \quad ...(3)$$

The equations (1) and (3) should represent the same line. So comparing the coefficients of 1/r, cos θ and sin θ, we get

$$1 = \frac{e + \cos\alpha}{A} = \frac{\sin\alpha}{B}$$

$$\Rightarrow \quad \cos \alpha = A - e \text{ and } \sin \alpha = B.$$

Squaring and adding, we have $(A - e)^2 + B^2 = 1$.

Which is the required condition.

Example 2:

A chord of a conic subtends a constant angle at a focus of the conic. Show that the chord touches another conic.

Solution:

Referred to the focus S as the pole, let the equation of the conic be

$$l/r = 1 + e \cos \theta. \quad ...(1)$$

Suppose a chord PQ of the conic (1) subtends a constant angle 2β at the focus S. Let α – β and α + β be the vectorial angles of the extremities of the chord PQ. Then the equation of the chord PQ is

$$l/r = e \cos \theta + \sec \beta \cos (\theta - \alpha)$$

$$\Rightarrow \quad (l \cos \beta)/r = e \cos \beta \cos \theta + \cos (\theta - \alpha). \quad ...(2)$$

Obviously the straight line (2) is the tangent to the conic $(l \cos \beta)/r = 1 + (e \cos \beta) \cos \theta$ at the point whose vectorial angle is a. Hence the proposition.

AUXILIARY CIRCLE

Definition: *The locus of the foot of the perpendicular from the focus on any tangent to a conic (ellipse or hyperbola) is a circle called the auxiliary circle of the conic.*

The equation the auxiliary circle. *To find the locus of the foot of the perpendicular from the focus of the conic $l/r = 1 + e \cos \theta$ on a tangent to it.*

or

To find the polar equation of the auxiliary circle of the conic

$$l/r = 1 + e \cos \theta.$$

The equation of the conic is given by

$$l/r = 1 + e \cos \theta. \quad ...(1)$$

Consider a point 'α' on (1). The equation of the tangent at the point 'α' is given by

$$l/r = \cos (\theta - \alpha) + e \cos \theta. \quad ...(2)$$

Changing (2) to cartesian coordinates, we get

$$l = (\cos \alpha + e)\, x + \sin \alpha\, y. \quad ...(2')$$

The equation of the line perpendicular to (2') and passing through the focus (*i.e.*, the pole or origin) is

$$0 = \sin \alpha.\, x - (\cos \alpha + e)\, y.$$

Changing it to polars, we obtain

$$0 = \sin \alpha.\, r \cos \theta - (\cos \alpha + e)\, r \sin \theta$$

$$\text{or} \quad \sin (\theta - \alpha) = - e \sin \theta. \quad ...(3)$$

Now the foot of the perpendicular from the focus S to the tangent (2) is given by the intersection of (2) and (3), and hence its locus is obtained by eliminating the variable 'α' between (2) and (3). The equation (2) and (3) may be rewritten as

$$\frac{l}{r} - e \cos \theta = \cos (\theta - \alpha)$$

and $- e \sin \theta = \sin (\theta - \alpha)$.

Squaring both side and adding these equations, we have

$$\left(\frac{l}{r} - e \cos\theta\right)^2 + e^2 \sin^2 \theta = 1,$$

$$\Rightarrow \qquad \frac{l^2}{r^2} - 2\frac{le}{r} \cos \theta + e^2 - 1 = 0$$

$$\Rightarrow \qquad \mathbf{(e^2 - 1)\, r^2 - 2ler \cos \theta = l^2 = 0.} \qquad \text{...(4)}$$

Which is the required equation of the auxiliary circle.

Particular case. If the conic be a parabola *i.e.*, $e = 1$, then the equation (4) becomes

$$-2ler \cos \theta + l^2 = 0$$

or $\quad l/r = 2 \cos \theta, \quad$ or $\quad l/r = \cos (\theta - 0) + 1. \cos \theta$

which is the equation of the tangnet to the parabola $l/r = 1 + \cos \theta$ at the vertex (*i.e.*, at the point $\theta = 0$).

TO FIND THE POINT OF INTERSECTION OF THE TWO TANGENTS AT THE POINTS A AND B ON THE CONIC

$$l/r = 1 + e \cos \theta.$$

The equation of the conic is $l/r = 1 + e \cos \theta$. ...(1)

The equations of the tangents to (1) at the points α and β are given as

$$l/r = \cos (\theta - \alpha) + e \cos \theta, \qquad \text{...(2)}$$

and $l/r = \cos (\theta - \beta) + e \cos \theta$, ...(3)

To find the points of intersection of (2) and (3), subtracting (3) from (2) we have

$$0 = \cos (\theta - \alpha) - \cos (\theta - \beta), \quad \text{or} \quad \cos (\theta - \alpha) = \cos (\theta - \beta).$$

$$\therefore \qquad \theta - \alpha = \pm (\theta - \beta)..$$

If we take the + ive sign, we get $\alpha = \beta$ which is inadmissible. So taking the – ive sign, we get

$$\theta - \alpha = - \theta + \beta, \quad \text{or} \quad \theta = \frac{1}{2} (\alpha + \beta). \qquad \text{...(4)}$$

Putting the value of θ from (4) in (2) or (3), we obtain

$$l/r = \cos \{\frac{1}{2}(\alpha + \beta) - \alpha\} + e \cos (\alpha + \beta)$$

$$= \cos \frac{1}{2}(\alpha - \beta) + e \cos \frac{1}{2}(\alpha + \beta).$$

If the point of intersection is (r', θ'), then we obtain

$$\theta' = \frac{1}{2}(\alpha + \beta) \text{ and } l/r' = \cos \frac{1}{2}(\alpha - \beta) + e \cos \frac{1}{2}(\alpha + \beta). \quad ...(5)$$

Particular case. If the conic is a parabola *l.e.*, e = 1, then from equation (5) we have

$$l/r' \cos \frac{1}{2}(\alpha - \beta) + \cos \frac{1}{2}(\alpha + \beta) = 2 \cos \frac{1}{2} \alpha \cos \frac{1}{2} \beta$$

$$\Rightarrow \quad r' = (l/2) \sec \frac{1}{2} \alpha \sec \frac{1}{2}\beta, \text{ and } \theta' = \frac{1}{2}(\alpha + \beta). \quad ...(6)$$

DIRECTOR CIRCLE

Definition: *The locus of the point of intersection of two perpendicular tangents to conic, is called the director circle of the conic.*

The equation of the director circle. To find the equation of the director circle of the conic l/r = 1 + e cos θ.

To find the equation of the conic is given as follows

$$l/r = 1 + e \cos \theta. \quad ...(1)$$

The director circle of the conic (1) is the locus of the point of intersection of perpendicular tangents to the conic (1).

The equations of the tangents to (1) at the points α and β are given by

$$l/r = \cos (\theta - \alpha) + e \cos \theta, \quad ...(2)$$

$$\text{and } l/r = \cos (\theta - \beta) + e \cos \theta, \quad ...(3)$$

From (2) and (3) we have

$$0 = \cos (\theta - \alpha) - \cos (\theta - \beta), \text{ or } \cos (\theta - \alpha) = \cos (\theta - \beta).$$

$$\therefore \quad \theta - \alpha = -(\theta - \beta) \text{ or } \theta = \frac{1}{2}(\alpha + \beta).$$

Putting $\theta = \frac{1}{2}(\alpha + \beta)$ in (2), we have

$$l/r = \cos \{\frac{1}{2}(\alpha + \beta) - \alpha\} + e \cos (\alpha + \beta)$$

$$= \cos \frac{1}{2}(\alpha - \beta) + e \cos \frac{1}{2}(\alpha + \beta).$$

Therefore if (r', θ') be the point of intersection of the tangents (2) and (3), then we have

$$\theta' = \frac{1}{2}(\alpha + \beta) \text{ and } l/r' = \cos\frac{1}{2}(\alpha - \beta) + e\cos\frac{1}{2}(\alpha + \beta) \quad \text{....(4)}$$

Changing the equation (2) of the tangent at the point α to cartesian form

$$l = (\cos\alpha + e)\,x + (\sin\alpha)\,y.$$

$\therefore$ m_1 = slope of the tangent (2) = $-(\cos\alpha + e)/(\sin\alpha)$.

Similarly m_2 = slope of the tangent (3) = $-(\cos\beta + e)/(\sin\beta)$.

The tangents (2) and (3) are perpendicular, if

$$m_1 m_2 = -1$$

$$\Rightarrow \quad -\frac{(\cos\alpha + e)}{\sin\alpha} \times -\frac{(\cos\beta + e)}{\sin\beta} = -1$$

$$\Rightarrow \quad (\cos\alpha\cos\beta + \sin\alpha\sin\beta) + e(\cos\alpha + \cos\beta) + e^2 = 0$$

$$\Rightarrow \quad \cos(a - b) + 2e\cos\left(\frac{\alpha+\beta}{2}\right)\cos\left(\frac{\alpha-\beta}{2}\right) + e^2 = 0$$

$$\Rightarrow \quad 2\cos^2\left(\frac{\alpha-\beta}{2}\right) - 1 + 2e\cos\left(\frac{\alpha+\beta}{2}\right)\cos\left(\frac{\alpha-\beta}{2}\right) + e^2 = 0 \quad \text{...(5)}$$

Now from (4)

$$\frac{1}{2}(\alpha + \beta) = \theta' \text{ and } \cos\frac{1}{2}(\alpha - \beta) = l/r' - e\cos\theta'. \quad \text{...(6)}$$

Eliminating α and β with the help of (5) and (6),

$$2\left(\frac{l}{r'} - e\cos\theta'\right)^2 + 2e\cos\theta' \cdot \left(\frac{l}{r'} - e\cos\theta'\right) + e^2 = 0$$

$$\Rightarrow \quad (1 - e^2)\,r'^2 + 2ler'\cos\theta' - 2l^2 = 0.$$

$\therefore$ the locus of (r', θ') is

$$(1 - e^2)\,r^2 + 2ler'\cos\theta - 2l^2 = 0. \quad \text{...(7)}$$

which is the required equation of the director circle.

Particular case. If the conic is a parabola *i.e.*, if e = 1, the equation (7) becomes as

$$2lr\cos\theta - l^2 = 0 \quad \text{or} \quad l/r = \cos\theta,$$

which is the equation of the directrix of the parabola $l/r = 1 + \cos\theta$.

Hence in the case of a parabola the locus of the point of intersection of perpendicular tangents is the directrix of the parabola.

Example:

If A, B C be any three point on a parabola, and the tangents at these point form a triangle A' B' C', show that SA SB.SC = SA'.SB'.SC', S being the focus of the parabola.

Solution:

Let the equation of the parabola be $l/r = 1 + \cos\theta$, referred to the focus S as the pole.

Let the vectorial angles of A, B, C be α, β, γ respectively. The equations of the tangents at these points are

$$l/r = \cos(\theta - \alpha) + \cos\theta, \qquad ...(1)$$

$$l/r = \cos(\theta - \beta) + \cos\theta, \qquad ...(2)$$

and $l/r = \cos(\theta - \gamma) + \cos\theta$, ...(3)

If C' is the point of intersection of the tangents (1) and (2), then the vectorial angle of C' $= \frac{1}{2}(\alpha = \beta)$ and the radius vector of C'

i.e., $$SC' = (l/2)\sec\frac{1}{2}\alpha\sec\frac{1}{2}\beta.$$

Similarly $$SA' = (l/2)\sec\frac{1}{2}\beta\sec\frac{1}{2}\gamma$$

and $$SB' = (l/2)\sec\frac{1}{2}\gamma\sec\frac{1}{2}\alpha.$$

$\therefore$ $$SA'.\ SB.\ SC' = (l^3/8)\sec^2\frac{1}{2}\alpha\sec^2\frac{1}{2}\beta\sec^2\frac{1}{2}\gamma \qquad ...(4)$$

Again since the point A whose vectorial angle is α and radius vector is SA lies on (1), therefore

$$l/SA = 1 + \cos\alpha = 2\cos^2\frac{1}{2}\alpha$$

or $$SA = (l/2)\sec^2\frac{1}{2}\alpha.$$

Similarly $SB = (l/2)\sec^2\frac{1}{2}\beta$

and $SC = (l/2)\sec^2\frac{1}{2}\gamma$.

$\therefore \quad SA.\ SB.SC = (l^3/8) \sec^2 \frac{1}{2} \alpha \sec^2 \frac{1}{2} \beta \sec^2 \frac{1}{2} \gamma. \qquad ...(5)$

From (4) and (5) the required result follows.

PAIR OF TANGENTS

To prove that the equation of the pair of tangents drawn to the conic $l/r = 1 + e \cos \theta$ from the point (r', θ') is

$$(S^2 - 1)(S'^2 - 1) = P^2,$$

where $\quad S \equiv l/r - e \cos \theta,\ S' \equiv l/r' - e \cos \theta'$

and $P \equiv (l/r - e \cos \theta)(l/r' - e \cos \theta') - \cos(\theta - \theta')$.

Hence to prove that the equations of the asymptotes of the conic are

$$le/r = (e^2 - 1) \cos \theta \pm \sqrt{(e^2 - 1)} \sin \theta.$$

Let as consider a point P with vectorial angle α on the given conic

$$l/r = 1 + e \cos \theta \qquad ...(1)$$

The tangent to (1) at the point a is given by

$$l/r = \cos(\theta - \alpha) + e \cos \theta. \qquad ...(2)$$

If the tangent (2) passes through the point (r', θ'), then we have

$$l/r' = \cos(\theta' - \alpha) + e \cos \theta. \qquad ...(3)$$

The required equation of the pair of tangents is obtained by eliminating α between (2) and (3). We have $(S^2 - 1)(S'^2 - 1)$

$\Rightarrow \quad \{l/r - e \cos \theta)^2 - 1\}\{l/r' - e \cos \theta')^2 - 1\}$

$\Rightarrow \quad \{\cos^2(\theta - \alpha) - 1\}\{\cos^2(\theta' - \alpha) - 1\}, \qquad$ (using (2) and (3))

$\Rightarrow \quad \{-\sin^2(\theta - \alpha)\}\{-\sin^2(\theta' - \alpha)\}$

$\Rightarrow \quad \sin^2(\theta - \alpha) \sin^2(\theta' - \alpha). \qquad ...(4)$

Also P

$\Rightarrow \quad (l/r - e \cos \theta)(l/r' - e \cos \theta') - \cos(\theta - \theta')$

$\Rightarrow \quad \cos(\theta - \alpha) \cos(\theta' - \alpha) - \cos(\theta - \theta') \qquad$ [using (2) and (3)]

$\Rightarrow \quad \frac{1}{2}\{2 \cos(\theta - \alpha) \cos(\theta' - \alpha)\} - \cos(\theta - \theta')$

$\Rightarrow \quad \frac{1}{2}\{\cos(\theta + \theta' - 2\alpha) + \cos(\theta - \theta')\} - \cos(\theta - \theta')$

$$\Rightarrow \quad \frac{1}{2}\{\cos(\theta + \theta' - 2\alpha) - \cos(\theta - \theta')\}$$

$$\Rightarrow \quad \frac{1}{2}\{2\sin(\theta - \alpha)\sin(\alpha - \theta')\} = -\sin(\theta - \alpha)\sin(\theta' - \alpha).$$

$$\therefore \quad P^2 = \sin^2(\theta - \alpha)\sin^2(\theta' - \alpha). \qquad ...(5)$$

From (4) and (5), we have

$$(S^2 - 1)(S'^2 - 1) = P^2. \qquad ...(6)$$

Since the equation (6) does not contain α, therefore it is the required equation of the pair of tangents drawn from the point (r', θ') to the conic

$$l/r = 1 + e\cos\theta.$$

TO FIND THE ASYMPTOTES

The asymptotes can be regarded as the pair of tangents drawn from the centre of the conic. The coordinates of the centre referred to the focus S as pole are

$$(ae, \pi) \text{ i.e., } \left(\frac{el}{1 - e^2}, \pi\right)$$

To find the asymptotes the point (r', θ') is to be taken as the point $\left(\frac{el}{1 - e^2}, \pi\right)$.

$$\therefore \quad r' = \frac{el}{1 - e^2} \quad \text{and} \quad \theta' = \pi \qquad ...(7)$$

Taking these values of r' and θ', then we have

$$\cos(\theta - \theta') = \cos(\theta - \pi) = -\cos\theta,$$

$$\left.\begin{aligned} S' &= \frac{l}{r'} - e\cos\theta' = \frac{1 - e^2}{e} - e\cos\pi \, \frac{1 - e^2}{e} + e = \frac{1}{e}, \\ \text{and } P &= \left(\frac{l}{r'} - e\cos\theta\right)\frac{1}{e} + \cos\theta = \frac{1}{er}. \end{aligned}\right\} \qquad ...(8)$$

Putting the values from (8) in (6), the equation of the asymptotes is given by

$$\left\{\left(\frac{l}{r'} - e\cos\theta\right)^2 - 1\right\}\left\{\frac{1}{e^2} - 1\right\} = \left(\frac{l}{er}\right)^2$$

$$\Rightarrow \left\{\frac{l^2}{r^2} - \frac{2le}{r}\cos\theta + e^2 + \cos^2\theta - 1\right\}\left(1 - e^2\right) = \frac{l^2}{r^2}$$

$$\Rightarrow \frac{l^2}{r^2} - \frac{e^2 l^2}{r^2} - \frac{2le}{r}(1 - e^2)\cos\theta + e^2(1 - e^2)\cos\theta - (1 - e^2) = \frac{l^2}{r^2}$$

$$\Rightarrow \frac{e^2 l^2}{r^2} + \frac{2le}{r}(1 - e^2)\cos\theta + e^2(1 - e^2)\cos^2\theta - (1 - e^2)$$

by transposition of terms.

Adding the term $(1 - e^2)\cos^2\theta$ to both sides, we obtain

$$\frac{e^2 l^2}{r^2} + \frac{2le}{r}(1 - e^2)\cos\theta + (1 - e^2)^2\cos^2\theta$$

$$= (1 - e^2)^2\cos^2\theta + e^2(1 - e^2)\cos^2\theta = (1 - e^2)$$

$$\Rightarrow \left\{\frac{el}{r} + \left(1 - e^2\right)\cos\theta\right\}^2 = (1 - e^2)\cos^2\theta\,\{1 - e^2 + e^2\} - (1 - e^2)$$

$$= (1 - e^2)(\cos^2\theta - 1) = -(1 - e^2)\sin^2\theta = (e^2 - 1)\sin^2$$

θ

Taking square root of both sides, we obtain

$$(le/r) + (1 - e^2)\cos\theta = \pm\sqrt{(e^2 - 1)}\sin\theta$$

$$\Rightarrow \quad le/r = (e^2 - 1)\cos\theta = \pm\sqrt{(e^2 - 1)}\sin\theta.$$

These are the equations of the asymptotes of the conic

CHORD OF CONTACT

To find the polar equation of the chord of contact of the point T (r', θ') with respect to the conic

$$l/r = 1 + e\cos\theta.$$

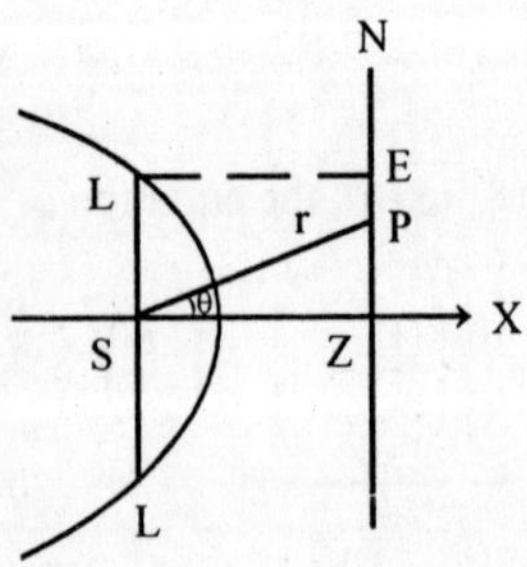

The equation. of the given conic is

$$l/r = 1 + e \cos \theta. \quad \text{...(1)}$$

Let P and Q be the points of contact of the tangents drawn from the point T (r', θ') to the conic (1).

The chord PD in the chord of contact of the point T with respect to the conic T with respect to the conic (1).

Let the vectorial angles of P and Q be α and β respectively.

The point of intersection T (r', θ') of the tangents at P and Q is given by

$$\theta' = \frac{1}{2}(\alpha + \beta), \quad \text{...(2)}$$

$$\text{and } l/r' = \cos \frac{1}{2}(\alpha - \beta) + e \cos \frac{1}{2}(\alpha + \beta). \quad \text{...(3)}$$

The equation of the chord PQ joining the points α and β is given by

$$l/r = \sec \frac{1}{2}(\alpha - \beta) \cos \{\theta - \frac{1}{2}(\alpha + \beta)\} + e \cos \theta. \quad \text{...(4)}$$

The equation (4) will become the equation of the chord of contact if α, β are eliminated with the help of (2) and (3).

The equation (4) may be written as

$$l/r - e \cos \theta = \frac{\cos\left\{\theta - \frac{1}{2}(\alpha + \beta)\right\}}{\cos \frac{1}{2}(\alpha - \beta)}$$

$$\Rightarrow \quad l/r - e \cos \theta = \frac{\cos(\theta - \theta')}{(l/r' - e \cos\theta')}$$

$$\Rightarrow \quad (l/r - e \cos \theta)(l/r - e \cos \theta') = \cos(\theta - \theta').$$

This is the required equation of the chord of contact of the point (r', θ') with respect to the conic

$$l/r = 1 + e \cos \theta.$$

POLAR

To find the equation of the polar of a point (r', θ') w.r.t. the conic

$$l/r = 1 + e \cos \theta.$$

We have to find the equation of the polar of a given point R (r', θ') with respect to the conic $l/r = 1 + e \cos \theta$.

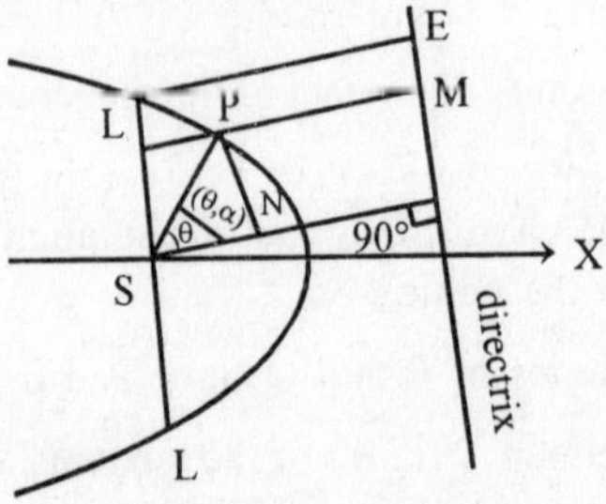

If we draw chords of the given conic passing through the point R, then the locus of the point of intersection of the tangents at the extremities of these chords is said to be the polar of R with respect to the given conic.

Let PQ be any such chord meeting the conic in points P and Q whose vectorial angles are α and β respectively.

If $P(r_1, \theta_1)$ is the point of intersection of the tangents at the points P 'α' and Q 'β', then

$$\theta_1 = \frac{1}{2}(\alpha + \beta), \qquad \text{...(1)}$$

and $l/r_1 = \cos \frac{1}{2}(\alpha - \beta) + e \cos \frac{1}{2}(\alpha + \beta)$

i.e., $l/r_1 - e \cos \theta_1 = \cos \frac{1}{2}(\alpha - \beta)$. ...(2)

The equation of the chord PQ joining the points a and b is given by

$$l/r' = \sec \frac{1}{2}(\alpha - \beta) \cos \{\theta - \frac{1}{2}(\alpha + \beta)\} + e \cos \theta.$$

Since it passes through the point R(r', θ'), therefore we have

$$l/r' = \sec \frac{1}{2}(\alpha - \beta) \cos \{\theta' - \frac{1}{2}(\alpha + \beta)\} + e \cos \theta'$$

$$\Rightarrow \quad l/r' - e \cos \theta' = \frac{\cos\left\{\theta' - \frac{1}{2}(\alpha + \beta)\right\}}{\cos \frac{1}{2}(\alpha - \beta)}$$

$$\Rightarrow \quad (l/r' - e \cos \theta') = \frac{\cos(\theta' - \theta_1)}{(l/r_1 - e \cos \theta_1)}, \text{ using (1) and (2)}$$

$$\Rightarrow \quad (l/r_1 - e \cos \theta_1)(l/r' - e \cos \theta') = \cos(\theta' - \theta_1). \qquad \text{...(3)}$$

$\therefore$ the polar of the point (r', θ') *i.e.,* the locus of the point (r_1, θ_1) is given by

$$(l/r_1 - e \cos \theta)(l/r' - e \cos \theta') = \cos(\theta' - \theta)$$

[(replacing r_1 by r and θ_1 by θ in (3)]

$$\Rightarrow \quad (l/r - e \cos \theta)(l/r' - e \cos \theta') = \cos(\theta - \theta'). \quad ...(4)$$

PERPENDICULAR LINES

Let the equation of a straight line by given as

$$l/r = A \cos \theta + B \sin \theta. \quad ...(1)$$

Multiplying both sides by r the equation (1) may be written as

$$l = Ar \cos \theta + Br \sin \theta.$$

Changing to cartesians this equation becomes

$$l = Ax + By \quad ...(2)$$

The equation of any line perpendicular to the line (2) is given by Bx – Ay + L where L is any real number

$$\Rightarrow \quad Br \cos \theta - Ar \sin \theta = L, \quad \text{(Changing to polars)}$$

$$\Rightarrow \quad L/r = A \cos\left(\frac{1}{2}\pi + \theta\right) + B \sin\left(\frac{1}{2}\pi + \theta\right). \quad ...(3)$$

Thus (3) is the equation of any line which is perpendicular to the line (1). In the equation (3) L is any real number.

Rule : The equation of any line perpendicular to

$$l/r = A \cos \theta + B \sin \theta$$

is obtained by writing $\theta + \frac{1}{2}\pi$ for θ and changing l to a new constant, say, L.

NORMAL

To find the equation of the normal at a point 'α' on the conic $l/r = 1 + e \cos \theta$.

Let $P(r_1, \alpha)$ be a point on the conic

$$l/r = 1 + e \cos \theta. \quad ...(1)$$

The tangent to (1) at the point $P(r_1, \alpha)$ is given by

$$l/r = \cos(\theta - \alpha) + e \cos \theta. \quad ...(2)$$

The normal to (1) at the point P is the line perpendicular to the line (2) and passing through the point P.

The equation of any line perpendicular to the line (2) is

$$L/r = \cos\left(\theta + \frac{1}{2}\pi - \alpha\right) + e\cos\left(\theta + \frac{1}{2}\pi\right)$$

$$\Rightarrow \quad L/r = -\sin(\theta - \alpha) - e\sin\theta. \quad ...(3)$$

The line (3) will be the normal at the point P if it passes through the point P (r_1, α).

So putting $r = r_1$ and $\theta = \alpha$ in (3), we have

$$L/r_1 = -\sin(\alpha - \alpha) - e\sin\alpha = -e\sin\alpha$$

$$\Rightarrow \quad L = -r_1 e\sin\alpha.$$

Now putting $L = -r_1 e\sin\alpha$ in (3), the equation of the normal to (1) at the point $P(r_1, \alpha)$ is

$$-(r_1 e\sin\alpha)/r = -\sin(\theta - \alpha) - e\sin\theta$$

$$\Rightarrow \quad \frac{er_1\sin\alpha}{r} = \sin(\theta - \alpha) + e\sin\theta \quad ...(4)$$

Again the point $P(r_1, \alpha)$ lies on (1).

$\therefore \quad l/r_1 = 1 + e\cos\alpha$ *i.e.*, $r_1 = l/(1 + e\cos\alpha)$.

Putting the value of r_1 in (4), the equation of the normal at the point P in terms of α alone, is given by

$$\frac{\mathbf{le\sin\alpha}}{(\mathbf{1 + e\cos\alpha})} \cdot \frac{\mathbf{1}}{\mathbf{r}} = \mathbf{\sin(\theta - \alpha) + e\sin\theta}. \quad ...(5)$$

SOLVED EXAMPLES

Example 1:

If the normals at α, β γ, δ on the conic

$$l/r = 1 + e\cos\theta$$

meet at a point (ρ, ϕ) (i.e., these normals are concurrent), prove that

$$\tan\frac{\alpha}{2}\tan\frac{\beta}{2}\tan\frac{\gamma}{2}\tan\frac{\delta}{2} + \left(\frac{1+e}{1-e}\right)^2 = 0,$$

and $\alpha + \beta + \gamma + \delta - 2\phi = (2n + 1)\pi$ i.e., an odd multiple of π radians.

Solution:

The given conic is

$$l/r = 1 + e \cos \theta. \quad ...(1)$$

The equation of the normal to (1) at any point 'λ' on it is

$$\frac{le \sin \lambda}{(1 + e \cos \lambda)} \cdot \frac{1}{r} = \sin (\theta - \lambda) + e \sin \theta.$$

If it passes through the point (ρ, ϕ), we have

$$le \sin \lambda = \rho\,(1+ e \cos \lambda)\,[\sin \phi \cos \lambda - \cos \phi \sin \lambda + e \sin \phi]. \quad ...(2)$$

Now putting

$$\sin \lambda = \frac{2t}{1 + t^2} \text{ and } \cos \lambda = \frac{1 - t^2}{1 + t^2}, \text{ where } t = \tan \frac{1}{2}\lambda,$$

the equation (2) becomes

$$\frac{le\, 2t}{1 + t^2} = \rho \left\{1 + \frac{e\left(1 - t^2\right)}{1 + t^2}\right\} \left\{\sin \phi \frac{1 - t^2}{1 + t^2} - \cos \phi \frac{2t}{1 + t^2} + e \sin \phi\right\}$$

$$\Rightarrow 2et(1 + t^2) = \rho[1 + e + t^2(1 - e)]\,[\sin \phi(1 + e) - 2t \cos \phi - t^2 \sin \phi\,(1 - e)]$$

$$\Rightarrow \rho\,(1 - e)^2 \sin \phi\, t^4 + 2\,\{le + (1 - e)\,\rho \cos \phi\}\, t^3 + 0.t^2$$

$$+ 2\,\{le + (1 + e)\,\rho \cos \phi\}\, t - \rho\,(1 + e^2) \sin \phi = 0. \quad ...(3)$$

The equation (3) being of degree 4 in *t* gives four values of *t i.e.,* of $\tan \frac{1}{2}\lambda$. Thus there are four points on the conic say, α, β, γ, and δ, the normals at which pass through the point (ρ, ϕ).

Let t_1, t_2, t_3, t_4, be the roots of the equation (3) where

$$t_1 = \tan \frac{1}{2}\alpha,\ t_2 = \tan \frac{1}{2}\beta,\ t_3 = \tan \frac{1}{2}\gamma,\ t_4 = \tan \frac{1}{2}\delta. \text{ then}$$

$$s_? = \Sigma t_1 = \frac{2\left\{le + (1 - e)\,\rho \cos \phi\right\}}{\rho\,(1 - e)^2 \sin \phi}$$

$$s_2 = \Sigma t_1 t_2 = 0, \quad [\because \text{ the coeff. of } t^2 \text{ in (3)} = 0]$$

$$s_3 = \Sigma t_1 t_2 t_3 = \frac{-2\{le + (1-e)\rho\cos\phi\}}{\rho(1-e)^2\sin\phi},$$

$$\text{and } s_4 = \Sigma t_1 t_2 t_3 t_4 = \frac{-\rho(1+e)^2\sin\phi}{\rho(1-e)^2\sin\phi} = -\frac{(1+e)^2}{(1-e)^2}$$

The last relation may be written as

$$\tan\frac{1}{2}\alpha\tan\frac{1}{2}\beta\tan\frac{1}{2}\gamma\tan\frac{1}{2}\delta = -\frac{(1+e)^2}{(1-e)^2},$$

This proves the first part of the question. From trigonometry, we have

$$\tan\left(\frac{\alpha}{2}+\frac{\beta}{2}+\frac{\gamma}{2}+\frac{\delta}{2}\right) = \frac{s_1 - s_3}{1 - s_2 + s_4}$$

$$= \frac{\dfrac{-2\{le+(1-e)\rho\cos\phi\}}{\rho(1-e)^2\sin\phi} + \dfrac{2\{le+(1+e)\rho\cos\phi\}}{\rho(1-e)^2\sin\phi}}{1 - 0 - \dfrac{(1+e)^2}{(1-e)^2}}$$

$$= \frac{2\times 2\rho e\cos\phi}{\rho(1-e)^2\sin\phi}\cdot\frac{(1+e)^2}{-4e} = -\cot\phi = \tan\left(\frac{\pi}{2}+\phi\right)$$

$$\text{Thus } \tan\left(\frac{\alpha}{2}+\frac{\beta}{2}+\frac{\gamma}{2}+\frac{\delta}{2}\right) = \tan\left(\frac{\pi}{2}+\phi\right).$$

$$\therefore \quad \frac{\alpha+\beta+\gamma+\delta}{2} = n\pi + \frac{\pi}{2} + \phi, \text{ where n is any interger}$$

$$\Rightarrow \alpha+\beta+\gamma+\delta-2\phi = (2n+1)\pi,$$

which is an odd multiple of π radians.

Example 2:

A chord PQ of a conic subtends a constant angle 2γ at the focus S and tangents at P and Q meet in T, prove that

$$\frac{1}{SP}+\frac{1}{SQ}-\frac{2\cos\gamma}{ST} = \frac{2\sin^2\gamma}{l}.$$

Solution:

Let the equation of the conic with focus S as the pole be

$$l/r = 1 + e \cos \theta. \quad ...(1)$$

If α, β are the vectorial angles of P, Q respectively and T (r', θ') be the point of intresection of the tangents at P and Q then proceeding as in § 9, we have

$$\theta' = \frac{\alpha + \beta}{2} \text{ and } \frac{l}{r'} = \cos \frac{\alpha - \beta}{2} + e \cos \frac{\alpha + \beta}{2} \quad ...(2)$$

Clearly $\alpha - \beta = 2\gamma$

and $r' = ST.$ ...(3)

Now the polar coordinates of the points P and Q are (SP, α) and (SQ, β) respectively. The points P and Q line on (1), therefore,

$$l/SP = 1 + e \cos \alpha,$$

and $l/SQ = 1 + e \cos \beta.$ ...(4)

Making use of the results (3) in (2), we have

$$l/ST = \cos \gamma + e \cos \frac{1}{2} (\alpha + \beta)$$

$$\Rightarrow \quad (2l \cos \gamma)/ST = 2 \cos^2 \gamma + 2e \cos \gamma \cos \frac{1}{2} (\alpha + \beta). \quad ...(5)$$

Adding both the relations given in (4) and subtracting (5), we get

$$\frac{l}{SP} - \frac{l}{SQ} - \frac{2l \cos \gamma}{ST} = 2 + e (\cos \alpha + \cos \beta) - 2 \cos^2 \gamma - 2e \cos \gamma \cos \frac{1}{2} (\alpha + \beta)$$

$$= 2 + 2e \cos \frac{1}{2} (\alpha + \beta) \cos \frac{1}{2} (\alpha - \beta) - 2 \cos^2 \gamma - 2e \cos \gamma \cos \frac{1}{2} (\alpha + \beta)$$

$$= 2 - 2 \cos^2 \gamma \qquad \left[\text{2nd and 4th terms cancel as } \gamma = \frac{1}{2} (\alpha + \beta)\right]$$

$$= 2 (1 - \cos^2 \gamma) = 2 \sin^2 \gamma.$$

$$\therefore \quad \frac{1}{SP} + \frac{1}{SQ} - \frac{2 \cos \gamma}{ST} = \frac{2}{l} \sin^2 \gamma.$$

Example 3:

If the tangent and normal at any point P of a conic meet the transverse axis in T and G respectively and if S be the focus, then show that $\frac{1}{SG} + \frac{1}{ST}$ *is constant.*

Solution:

Let the conic be

$$l/r = 1 + e \cos \theta,$$

the focus S being at the pole.

Let the vectorial angle of P be 'α'.

The equations of the tangent and normal to (1) at the point P are

$$l/r = \cos(\theta - \alpha) + e \cos \theta, \qquad ...(2)$$

and $$\frac{le \sin \alpha}{(1 + e \cos \alpha)} \frac{1}{r} = \sin(\theta - \alpha) + e \sin \theta \qquad ...(3)$$

respectively.

The tangent (2) meets the axis *i.e.*, the initial line in T where r = ST and θ = 0. Putting these values in the equation (2), we have

$$l/ST = \cos(0 - \alpha) + e \cos 0.$$

$$\therefore \quad l/ST = (e + \cos \alpha)/l. \qquad ...(4)$$

The normal (3) meets the axis in G where r = SG, and θ = π.

Putting these values in the equation (3), we have

$$\frac{le \sin \alpha}{(1 + e \cos \alpha)} \cdot \frac{1}{SG} = \sin(\pi - \alpha) + e \sin \pi$$

$$\therefore \quad \frac{1}{SG} = \frac{(1 + e \cos \alpha)}{le}. \qquad ...(5)$$

Subtracting (4) from (5), we get

$$\frac{1}{SG} - \frac{1}{ST} = \frac{1 + e \cos \alpha}{le} - \frac{e + \cos \alpha}{l} = \frac{1 - e^2}{le} = \text{constant.}$$

Example 4(a):

Prove that, if chords of a conic subtend a constant angle at a focus, the tangents at the ends of the chord will meet on a fixed conic and the chord will touch (or envelope) another fixed conic.

Solution:

Let the given conic by $l/r = 1 + e \cos \theta$, the focus S being at the pole. Suppose PQ is a variable chord of the given conic and let $\angle PSQ = 2\gamma$. Proceeding as in above, the equation of the locus of the point of intersection T of the tangents at P and Q is

$$(l \sec \gamma)/r = 1 + (e \sec \gamma) \cos \theta,$$

which is a fixed conic and this proves the first part of the question.

Now the equation of the chord PQ is

$$l/r = \sec \frac{1}{2} (\alpha - \beta) \cos \left\{\theta - \frac{1}{2}(\alpha + \beta)\right\} + e \cos \theta$$

$$\Rightarrow \quad l/r = \sec \gamma \cos \left\{\theta - \frac{1}{2}(\alpha + \beta)\right\} + e \cos \theta \quad [\because \alpha - \beta = 2\gamma]$$

$$\Rightarrow \quad (l \cos \gamma)/r = \cos \left\{\theta - \frac{1}{2}(\alpha + \beta)\right\} + (e \cos \gamma) \cos \theta$$

which clearly touch (or envelopes) the conic whose equation is

$$(l \cos \gamma)/r = 1 + (e \cos \gamma) \cos \theta.$$

This is a fixed conic because γ is a constant.

Example 4(b):

If the normals at three points of the parabola

$$r = a \operatorname{cosec}^2 \frac{1}{2} \theta$$

whose vectorial angles are α, β and γ meet in a point whose vectorial angle is ϕ, prove that $2\phi = \alpha + \beta + \gamma + \pi$.

Soluiton:

The given parabola is

$$r = a \operatorname{cosec}^2 \frac{1}{2} \theta$$

or $$2a/r = 2 \sin^2 \frac{1}{2} \theta$$

or $$2a/r = 1 - \cos \theta.$$

The normal to this parabola at the point 'λ' on it is

$$\frac{-2a \sin \lambda}{1 - \cos \lambda} \cdot \frac{1}{r} = \sin (\theta - \lambda) - \sin \theta.$$

If it passes through the points (ρ, φ), we have

$$\frac{-2a \sin \lambda}{1 - \cos \lambda} \cdot \frac{1}{\rho} = \sin (\theta - \lambda) - \sin \phi.$$

$$\Rightarrow \quad \frac{-2a.\ 2 \sin \frac{1}{2}\lambda \cos \frac{1}{2}\lambda}{2 \sin^2 \frac{1}{2}\lambda} = r\, [\sin \phi \cos \lambda - \cos \phi \sin \lambda - \sin \phi]$$

$$\Rightarrow \quad -2a \cot \frac{1}{2}\lambda = \rho\, [- \sin \phi\, (1 - \cos \lambda) - \cos \phi \sin \lambda]$$

$$\Rightarrow \quad 2a = \rho \tan \frac{1}{2}\lambda \left[\sin \phi . 2 \sin^2 \frac{1}{2}\lambda + 2 \cos \phi \sin \frac{1}{2}\lambda \cos \frac{1}{2}\lambda\right].$$

Dividing both sides by 2 cos² *1/2λ*, we have

$$a \sec^2 \frac{1}{2}\lambda = \rho \tan \frac{1}{2}\lambda \left[\sin \phi \tan^2 \frac{1}{2}\lambda + \cos \phi \tan \frac{1}{2}\lambda\right]$$

$$\Rightarrow \quad a\left(1 + \tan^2 \frac{1}{2}\lambda\right) = \rho \sin \phi \tan^3 \frac{1}{2}\lambda + \rho \cos \phi \tan \frac{1}{2}\lambda$$

$$\Rightarrow \quad \rho \sin \phi \tan^3 \frac{1}{2}\lambda + (\rho \cos \phi - a) \tan^2 \frac{1}{2}\lambda + 0 \tan \frac{1}{2}\lambda - a = 0....(1)$$

This equation being a cubic in tan $\frac{1}{2}\lambda$ gives values of tan $\frac{1}{2}\lambda$ and hence three values of λ. Thus there are three are three points on the parabola the normals at which pass through the point (ρ, φ). If these points are α, β, γ, then tan $\frac{1}{2}\alpha$ tan $\frac{1}{2}\beta$ tan $\frac{1}{2}\gamma$ are the roots of the cubic (1) tan $\frac{1}{2}\lambda$.

By the theory of equations, we have for the cubic (1),

$$\Sigma \tan \frac{1}{2}\alpha = -(\rho \cos \phi - a)/\rho \sin \phi,\ \Sigma \tan \frac{1}{2}\alpha \tan \frac{1}{2}\beta = 0,$$

$$\tan \frac{1}{2}\alpha \tan \frac{1}{2}\beta \tan \frac{1}{2}\gamma = a/\rho \sin \phi.$$

Now from trigonometry, we have

$$\tan\left(\frac{1}{2}\alpha + \frac{1}{2}\beta + \frac{1}{2}\gamma\right) = \frac{\sum \tan \frac{1}{2}\alpha - \tan \frac{1}{2}\alpha \tan \frac{1}{2}\beta \tan \frac{1}{2}\gamma}{1 - \sum \tan \frac{1}{2}\alpha \tan \frac{1}{2}\beta}$$

$$= \frac{\{-(\rho \cos \phi - a)/\rho \sin \phi\} - (a/\rho \sin \phi)}{1 - 0}$$

$$= -\frac{(\rho \cos \phi - a)}{\rho \sin \phi} - \frac{a}{\rho \sin \phi}$$

$$= \frac{-\rho \cos \phi + a - a}{\rho \sin \phi} = -\cot \phi$$

$$= \tan\left(\frac{1}{2}\pi + \phi\right)$$

$$\therefore \quad \frac{1}{2}\alpha + \frac{1}{2}\beta + \frac{1}{2}\gamma = \frac{1}{2}\pi + \phi$$

or

$$\alpha + \beta + \gamma = \pi + 2\phi$$

$$\Rightarrow \quad 2\phi = \alpha + \beta + \gamma - \pi.$$

Example 4(c):

Prove that the locus of the point of intersection of tangents at the extremities of perpendicular focal radii of a conic is another conic having the same focus.

Solution:

Suppose PS and QS are two perpendicular focal radii of the conic $l/r = 1 + e \cos \theta$, the focus S being at the pole. Let α and β be the vectorial angles of P and Q respectively. Then her $\alpha - \beta = \pi/2$ or if $\alpha - \beta = 2\gamma$, then $\gamma = \pi/4$.

Now proceeding as in above, the locus of the point of intersection T of the tangents at P and Q is

$$(l \sec \gamma)/r = 1 + (e \sec \gamma) \cos \theta$$

$$\Rightarrow \quad \left(l \sec \frac{1}{4}\pi\right)/r\ 1 + \left(e \sec \frac{1}{4}\pi\right) \cos \theta$$

$$\Rightarrow \quad (l \sqrt{2})/r = 1 + (e \sqrt{2}) \cos \theta.$$

This is the equation of a conic with focus at the pole. Thus this conic has the same focus as of the given conic.

Example 4(d):

If the normals at α, β, γ on the parabola $l/r = 1 + \cos \theta$ meet at a point (ρ, ϕ) prove that $2\phi = \alpha + \beta + \gamma$.

Solution:

Q '$\pi/2+\alpha$'

The equation of the normal to the parabola $l/r = 1 + \cos\theta$ at the point 'λ' on it is

$$\frac{l \sin\lambda}{1+\cos\lambda}\cdot\frac{1}{r} = \sin(\theta-\lambda)+\sin\theta.$$

If it passes through the point (ρ, ϕ), we have

$$\frac{l \sin\lambda}{1+\cos\lambda}\cdot\frac{1}{\rho} = \sin(\phi-\lambda)+\sin\phi.$$

Now proceeding as in this equation takes the form $l\tan^2\frac{1}{2}\lambda + 0.$

$$\tan^2\frac{1}{2}\lambda + (1+2\rho\cos\phi)\tan\frac{1}{2}\lambda - 2\rho\sin\phi = 0. \qquad ...(1)$$

This equation bieng a cubic in $\tan\frac{1}{2}\lambda$ gives three values of $\tan\frac{1}{2}\lambda$ and hence three values of λ. Thus there are three points on the parabola the normals at which pass through the point (ρ, ϕ). If these points are α, β, γ, then $\tan\frac{1}{2}\alpha$, $\tan\frac{1}{2}\beta$, $\tan\frac{1}{2}\gamma$, are the roots of the cubic (1) in $\tan\frac{1}{2}\lambda$.

By the theory of equations, we have for the cubic (1),

$$\tan\frac{1}{2}\alpha + \tan\frac{1}{2}\beta + \tan\frac{1}{2}\gamma = \sum\tan\frac{1}{2}\alpha = 0,$$

$$\Sigma\tan\frac{1}{2}\alpha\tan\frac{1}{2}\beta = (l+2\rho\cos\phi)/l,$$

$$\tan\frac{1}{2}\alpha\tan\frac{1}{2}\beta\tan\frac{1}{2}\gamma = 2\rho\sin\phi\, l.$$

Now from trigonometry, we have

$$\tan\left(\frac{1}{2}\alpha+\frac{1}{2}\beta+\frac{1}{2}\gamma\right) = \frac{\sum\tan\frac{1}{2}\alpha - \tan\frac{1}{2}\alpha\tan\frac{1}{2}\beta\tan\frac{1}{2}\gamma}{1-\sum\tan\frac{1}{2}\alpha\tan\frac{1}{2}\beta}$$

$$= \frac{0-(2\rho\sin\phi/1)}{0-\{(1+2\rho\sin\phi/1)\}} = \frac{-2\rho\sin\phi/1}{1-(1+2\rho\cos\phi/1)} = \tan\phi$$

$$\therefore \qquad \frac{1}{2}\alpha+\frac{1}{2}\beta+\frac{1}{2}\gamma = \phi$$

or $\qquad \alpha+\beta+\gamma = 2\phi.$

Example 5:

Prove that the radius vector TS of the point of intersection tangents at P and Q bisects the angle between the radii vectors of P and Q (i.e., between PS and QS).

Solution:

Let α, β be the vectorial angles of P and Q respectively.

If θ' is the vectorial angle of the point of intersection T of the tangents at P and Q, then

$$\theta' = \angle TSX = \frac{1}{2}(\alpha + \beta).$$

$$\therefore \qquad \angle TSQ = \angle TSX - \angle QSX$$

$$= \frac{1}{2}(\alpha - \beta) - \beta$$

$$= \frac{1}{2}(\alpha - \beta)$$

and $\angle PST = \angle PSX - \angle TSX$

$$= \alpha - \frac{1}{2}(\alpha - \beta) = \frac{1}{2}(\alpha - \beta).$$

$$\therefore \qquad \angle TSQ = \angle PST = \frac{1}{2}\angle PSQ.$$

This proves the required result.

Example 6:

Find the equation of the circle circumscribing the triangle formed by tangents at three given points of a parabola.

Solution:

Let that thre points on the parabola

$$l/r = 1 + \cos\theta$$

be A, B, C and let α, β, γ, be thir vectorial angles. Also let A', B', C' be the points of intersection of these tangents. Then as in above,

the point A' is $\left(\frac{l}{2}\sec\frac{\beta}{2}\sec\frac{\gamma}{2}, \frac{\beta + \gamma}{2}\right)$,

the point B' is $\left(\frac{l}{2}\sec\frac{\alpha}{2}\sec\frac{\gamma}{2}, \frac{\alpha + \gamma}{2}\right)$, and the point C' is $\left(\frac{l}{2}\sec\frac{\alpha}{2}\sec\frac{\beta}{2}, \frac{\alpha+\beta}{2}\right)$,

By actual substitution we see that the three points A', B', C' lie on the curve

$$r = \frac{l}{2}\sec\frac{\alpha}{2}\sec\frac{\beta}{2}\sec\frac{\gamma}{2}\cos\left(\theta - \frac{\alpha+\beta+\gamma}{2}\right). \qquad ...(1)$$

The equation (1) is of the type

$$r = 2a\cos(\theta - \lambda)$$

which is the equation of a circle passing through the pole, the diameter through the pole making an angle λ with the initial line and the length of the diameter equal to 2a.

Hence the equation (1) is the equation of the circumcircle of the triangle A' B' C', the length of the diameter of the circumcircle being equal to

$$(l/2)\sec\frac{1}{2}\alpha\sec\frac{1}{2}\beta\sec\frac{1}{2}\gamma$$

and the diameter through the pole making an angle $\frac{1}{2}(\alpha+\beta+\gamma)$ with the initial line.

Remark:

The centre of the circle (1) is the middle point of its diameter passing through the pole. So the vectorial angle of the centre of the circle (1) is $\frac{1}{2}(\alpha+\beta+\gamma)$ and the radius vector is

$$(l/4)\sec\frac{1}{2}\alpha\sec\frac{1}{2}\beta\sec\frac{1}{2}\gamma.$$

Example 7(a):

Prove that the centres of the four circles circumscribing the four triangles formed by the four tangents drawn to a parabola at points whose vectorial angles are α, β, γ, δ lie on another circle which passes through the focus of the parabola.

Solution:

Let α, β, γ and δ be the vectorial angles of the four points A, B, C and D respectively on the parabola

$$l/r = 1 + \cos\theta.$$

There are four tangents at the four points A, B, C, D of the parabola. To form a triangle, we require only three tangents (since a triangle has three sides). Three tangents out of four can be chosen in ${}^4C_3$ *i.e.,* 4 ways. Hence four triangles are formed by the four tangents.

Now from the equation (1) of [give complete proof here], the polar coordinates of the centre of the circle circumscribing the triangle formed by the tangents at A, B, C are

$$\left(\frac{l}{4}\sec\frac{\alpha}{2}\sec\frac{\beta}{2}\sec\frac{\gamma}{2}, \frac{\alpha+\beta+\gamma}{2}\right).$$

Similarly the coordinates of the centres of the other three circum-cricles are

$$\left(\frac{l}{4}\sec\frac{\alpha}{2}\sec\frac{\beta}{2}\sec\frac{\delta}{2}, \frac{\alpha+\beta+\delta}{2}\right), \text{ etc.}$$

These four points clearly lie on the circle woese equation may be put as

$$r = \frac{l}{4}\sec\frac{\alpha}{2}\sec\frac{\beta}{2}\sec\frac{\gamma}{2}\sec\frac{\delta}{2}\cos\left(\theta - \frac{\alpha+\beta+\gamma+\delta}{2}\right)$$

This is the equation of a circle passing through the pole *i.e.,* the focus of the given parabola.

Example 7(b):

PQ is a variable chord of a conic having S for focus and angle PSQ is constant. Prove that the locus of the point of intersection of tangents at P and Q is a conic having S for a focus and the corresponding directrix is common with the given conic.

Solution:

Let the given conic be

$$l/r = 1 + e\cos\theta, \qquad ...(1)$$

the focus S being at the pole.

Let α, β, be the vectorial angles of the points P and Q respectively.

Let T(r', θ') be the point of intersection of the tangents at P and Q. Then we have

$$\theta' = \frac{\alpha+\beta}{2}. \qquad ...(2)$$

and
$$\frac{1}{r'} = \cos\left(\frac{\alpha-\beta}{2}\right) + e\cos\left(\frac{\alpha-\beta}{2}\right). \qquad ...(3)$$

(Derive these results here)

Now if ∠PSQ = 2γ (say), where γ is a constant, then

$$\alpha - \beta = 2\gamma. \qquad ...(4)$$

The locus of T will be obtained by eliminating α and β between (2), (3) and (4). Substituting the values of $\alpha + \beta$ and $\alpha - \beta$ from (2) and (4) in (3), we have

$$l/r' = \cos\gamma + e\cos\theta'$$

$$\Rightarrow \quad (l\sec\gamma)/r' = 1 + (e\sec\gamma)\cos\theta'.$$

$\therefore$ the locus of T (r', θ') is

$$(l\sec\gamma)/r = 1 + (e\sec\gamma)\cos\theta,$$

which is a conic whose focus is the pole S, and the equation of its directrix corresponding to the focus S is

$$(l\sec\gamma)/r = (e\sec\gamma)\cos\theta$$

i.e., $l/r = e\cos\theta$

which is also the directrix of the given conic (1). Hence the proposition.

Example 7(c):

P, Q, R are three points on the conic $l/r = 1 + e\cos\theta$ the focus S being the pole. The tangent at Q meets SP and SR in M and N so that SM = SN = l.

Prove that the chord PR touches the conic

$$l/r = 1 + 2e\cos\theta.$$

Solution:

The equation of the conic with focus S as the pole is

$$l/r = 1 + 2e\cos\theta.$$

Suppose α, β, γ are the vectorial angles of the points P, Q, R respectively.

The equation of the chord PR joining the points α and γ is

$$(l/r = \sec\frac{1}{2}(\gamma - \alpha)\cos\left\{\theta - \frac{1}{2}(\gamma + \alpha)\right\} + e\cos\theta. \qquad ...(1)$$

It is given that the tangent at Q meets SP and SR in M and N respectively and SM = SN = l. Therefore the polar coordinates of M and N are (l, α) and (l, γ) respectively.

The tangent at Q is

$$l/r = \cos(\theta - \beta) + e\cos\theta. \qquad ...(2)$$

The points M (l, α) and N (l, γ) lie on (2). Therefore

$$1 = \cos(\alpha - \beta) + e \cos \alpha,$$

and $$1 = \cos(\gamma - \beta) - e \cos \gamma.$$

These equations may be written as

$$(\cos \beta + e) \cos \alpha + \sin \beta \sin \alpha - 1 = 0, \qquad \text{...(3)}$$

and $$(\cos \beta + e) \cos \gamma + \sin \beta \sin \gamma - 1 = 0. \qquad \text{...(4)}$$

Solving (3) and (4) for $\cos \beta + e$ and $\sin \beta$, we have

$$\frac{\cos \beta + e}{\sin \gamma - \sin \alpha} = \frac{\sin \beta}{\cos \alpha - \cos \gamma}$$

$$= \frac{1}{\cos \alpha \sin \gamma - \cos \gamma \sin \alpha} = \frac{1}{\sin(\gamma - \alpha)}$$

$$\Rightarrow \frac{\cos \beta + e}{2 \cos \frac{1}{2}(\gamma + \alpha) \sin \frac{1}{2}(\gamma + \alpha)} = \frac{\sin \beta}{2 \sin \frac{1}{2}(\gamma + \alpha) \sin \frac{1}{2}(\gamma + \alpha)}$$

$$= \frac{1}{2 \sin \frac{1}{2}(\gamma - \alpha) \cos \frac{1}{2}(\gamma - \alpha)}$$

$$\therefore \quad \cos \beta + e = \cos \frac{1}{2}(\gamma - \alpha) \cos \frac{1}{2}(\gamma - \alpha), \qquad \text{...(5)}$$

and $$\sin \beta = \sin \frac{1}{2}(\gamma - \alpha) \sin \frac{1}{2}(\gamma - \alpha). \qquad \text{...(6)}$$

Now the equation (1) of the chord PR may be re-written as

$$l/r = \sec \frac{1}{2}(\gamma - \alpha) \cos \theta \sec \frac{1}{2}(\gamma - \alpha)$$

$$+ \sec \frac{1}{2}(\gamma - \alpha) \sin \theta \sec \frac{1}{2}(\gamma - \alpha) + e \cos \theta$$

$\Rightarrow l/r = (\cos \beta + e) \cos \theta \sin \theta \sin \beta + e \cos \theta$ (using (5) and (6))

$$\Rightarrow l/r = \cos \theta \cos \beta + \sin \theta \sin \beta + 2e \cos \theta$$

$$\Rightarrow l/r \cos(\theta - \beta) + 2e \cos \theta$$

which is theequation of the tangent at the point 'β' to the coic $l/r = 1 + 2e \cos \theta$. Hence the chord PR touches the conic

$$l/r = 1 + 2e \cos \theta.$$

Example 8:

If thetangenst at any two points P and Q of a conic meet in a pointT, and ifthe chord PQ meets thedirctrix corresponding to S in a point K, prove that ∠KST isa right angle.

Solution:

Let the equation of the conic with focus S as the pole be

$$l/r = 1 + e \cos \theta. \quad ...(1)$$

Let α, β be the vectorial angles of the points P and Q on (1). the equations of the tangents at P and Q are

$$l/r \cos(\theta - \alpha) + e \cos \theta \quad ...(2)$$

and $$l/r \cos(\theta - \beta) + e \cos \theta. \quad ...(3)$$

The vectorial angle of the point of intersection T of the tangents (2) and (3) is

$$(\alpha + \beta)/2.$$

$$\therefore \quad \angle TSZ = (\alpha + \beta)/2.$$

the equation of the chord PQ is

$$l/r = \sec \frac{1}{2}(\alpha - \beta) \cos\left\{\theta - \frac{1}{2}(\alpha + \beta)\right\} + e \cos \theta. \quad ...(4)$$

The equation of the directrix ZK corresponding to the focus S is

$$l/r = e \cos \theta. \quad ...(5)$$

To find the vectorial angle of the point of intersection K of (4) and (5) putting the value of l/r from (5) in (4), we have

$$e \cos \theta = \sec \frac{1}{2}(\alpha - \beta) \cos\left\{\theta - \frac{1}{2}(\alpha + \beta)\right\} + e \cos \theta.$$

$$\Rightarrow \quad \sec \frac{1}{2}(\alpha - \beta) \cos\left\{\theta - \frac{1}{2}(\alpha + \beta)\right\} = 0$$

$$\Rightarrow \quad \cos\left\{\theta - \frac{1}{2}(\alpha + \beta)\right\} = 0 \quad \left[\because \sec \frac{1}{2}(\alpha - \beta) \neq 0\right]$$

$$\therefore \quad \theta - \frac{1}{2}(\alpha + \beta) = \pm \frac{1}{2}\pi.$$

But from the figure we observe that for the point K the vectorial angle θ is less then $\frac{1}{2}(\alpha + \beta)$. So for the point K, we have

$$\theta - \frac{1}{2}(\alpha + \beta) = -\frac{1}{2}\pi$$

$$\Rightarrow \qquad \theta = \frac{1}{2}(\alpha + \beta) = -\frac{1}{2}\pi \text{ i.e., } \angle KSZ = \frac{1}{2}(\alpha + \beta) - \frac{1}{2}\pi.$$

Now $\angle KST = \angle TSZ - \angle KSZ = \frac{1}{2}(\alpha + \beta) - \left\{\frac{1}{2}(\alpha + \beta) - \frac{1}{2}\pi\right\} = \frac{1}{2}\pi$

Therefore $\angle KST$ is a rigth angle.

Example 9:

PSP' is a focal chord of the conic. Prove that the tangents at P and P' intersect on the directrix.

Solution :

Let the equation of the conic be $l/r = 1 + e \cos \theta$.

Suppose the focal chord PSP' is inclined at an angle α to the initial line so that the vectorial angles of P and P' are α and $\pi + \alpha$ respectively. The tangents at P and P' are

$$l/r = \cos(\theta - \alpha) + e \cos \theta \qquad \text{...(1)}$$

and $l/r = \cos\{\theta - (\pi + \alpha)\} + e \cos \theta$

$$\Rightarrow \qquad l/r = -\cos(\theta - \alpha) + e \cos \theta. \qquad \text{...(2)}$$

The locus of the point of intersection of the tangents (1) and (2) is obtained by eliminating α between (1) and (2).

Adding (1) and (2), we get $2l/r = 2e \cos \theta$

i.e., $l/r = e \cos \theta$

which is the equation of the directrix. Hence the tangents (1) and (2) intersect on the directrix.

Example 10(a):

Two equal ellipses of eccentricity e, are placed with their axes at right angles and they have one focus S in common. If PQ be a common tangent, show that the angle PSQ is equal to 2 $\sin^{-1}\left(e/\sqrt{2}\right)$.

Solution :

Take the common focus S as the pole and the axis of one ellipse as the initial line so that the axis of the other ellipse makes angle $\frac{1}{2}\pi$ with the initial line. Let the equations to the two equal ellipses be

$$l/r = 1 + e \cos\theta \qquad ...(1)$$

and $l/r = 1 + e \cos(\theta - \pi/2)$,

$$\Rightarrow \qquad l/r = 1 + e \sin\theta. \qquad ...(2)$$

It is given that PQ is a common tangent to the two ellipses. Let the vectorial angles of P, a point on (1), and Q, a point on (2), be α and β respectively. Therefore the tangent to (1) at the point α, *i.e.,*

$$l/r = \cos(\theta - \alpha) + \cos\theta,$$

$$\Rightarrow \qquad l/r = (\cos\alpha + e)\cos\theta + \sin\alpha \sin\theta \qquad ...(3)$$

and the tangent to (2) at the point β *i.e.,*

$$l/r = \cos(\theta - \beta) + e \sin\theta$$

$$\Rightarrow \qquad l/r = \cos\beta \cos\theta + (\sin\beta + e)\sin\theta \qquad ...(4)$$

should be identical. Hence comparing (3) and (4), we have

$$1 = \frac{\cos\alpha + e}{\cos\beta} = \frac{\sin a}{\sin(\beta + e)}$$

$\therefore \quad \cos e + \alpha = \cos\beta$ *i.e.,* $\cos\beta - \alpha \cos\alpha = e$

and $\sin\alpha = \sin\beta + e$ *i.e.,* $\sin\alpha - \sin\beta = e$.

Squaring and adding, we get

$$2 - 2(\cos\alpha \cos\beta + \sin\alpha \sin\beta) = 2e^2$$

$$\Rightarrow \qquad \cos(\alpha - \beta) = 1 - e^2,$$

$$\text{or} \qquad 1 - 2\sin^2 \frac{1}{2}(\alpha - \beta) = 1 - e^2$$

$$\Rightarrow \qquad \sin^2 \frac{1}{2}(\alpha - \beta) = \frac{1}{2}e^2,$$

$$\text{or} \qquad \sin \frac{1}{2}(\alpha - \beta) = e/\sqrt{2}$$

$$\therefore \qquad \frac{1}{2}(\alpha - \beta) = \sin^{-1}\left(e/\sqrt{2}\right).$$

$$\therefore \qquad \angle PSQ = \alpha - \beta = 2\sin^{-1}\left(e/\sqrt{2}\right).$$

Example 10(b):

Show that the locus of the feet of perpendiculars from the focus S of a conic on chords subtending a constant angle 2γ at S is the circle whose polar equation referred to S as pole is

$$r^2 (e^2 - \sec^2 \gamma) - 2ler \cos \theta + l^2 = 0$$

where 2l is the latus rectum and e the eccentricity of the conic.

Solution:

Referred to the focus S as the pole let the equation of the conic be

$$l/r = 1 + e \cos \theta. \qquad ...(1)$$

Let PQ be a chord of (1) subtending an angle 2γ at the focus S. Let $\alpha - \gamma$ and $\alpha + \gamma$ be the vectorial angles of the extremities of the chord PQ so that $\angle QSP = 2\gamma$; here α is a parameter. Then the equation of the chord PQ is

$$l/r = e \cos \theta + \sec \gamma \cos (\theta - \alpha) \qquad ...(2)$$

$$\Rightarrow \quad l/r = e \cos \theta + \sec \gamma \cos \theta + \sec \gamma \sin \theta \sin \alpha$$

$$\Rightarrow l = (e + \sec \gamma \cos \alpha) r \cos \theta + (\sec \gamma \sin \alpha) r \sin \theta. \qquad (2')$$

Mentally transforming (2) to cartesians, we see that the equation of the perpendicular drawn from the focus S (*i.e.*, the pole or origin) to the line (2') is

$$0 = (e + \sec \gamma \cos \alpha) r \sin \theta - (\sec \gamma \sin \alpha) r \cos \theta$$

$$\Rightarrow \quad - e \sin \theta = \sec \gamma \sin (\theta - \alpha). \qquad ...(3)$$

The foot of the perpendicular drawn from the focus S to the chord (2) is the point of intersection of the lines (2) and (3). To find its locus we are to eliminate the variable α between (2) and (3).

The equation (2) can be written as

$$(l/r - e \cos \theta) = \sec \gamma \cos (\theta - \alpha). \qquad ...(4)$$

Squaring and adding (3) and (4), we get

$$e^2 \sin^2 \theta + (l/r - e \cos \theta)^2 = \sec^2 \gamma$$

$$\Rightarrow \quad e^2 + \frac{l^2}{r^2} - \frac{2le}{r} \cos \theta = \sec^2 \gamma$$

$$\Rightarrow \quad r^2 (e^2 - \sec^2 \gamma) - 2ler \cos \theta + l^2 = 0$$

which is the required locus.

Example 10(c):

If PSQ and PS'R be two chords of an ellipse through the foci S and S', show that $\frac{PS}{SQ} + \frac{PS'}{S'R}$ *is independent of the position of P.*

Solution :

Let the polar equation of the ellipse be given as

$l/r = 1 + e \cos \theta.$ ...(1)

Since PSQ is a focal chord, therefore if the vectorial angle of P is α then that of Q is $\pi + \alpha$.

$\therefore\ l/SP = 1 + e \cos \alpha,$ ...(2)

and $l/SQ = 1 + e \cos (\pi + \alpha) = 1 - e \cos \alpha.$ ...(2)

Adding (2) and (3), we obtain

$$\frac{1}{SP}+\frac{1}{SQ}=2, \text{ or } \frac{1}{SP}+\frac{1}{SQ}=\frac{2}{l} \qquad ...(4)$$

Multiplying both sides of (4) by SP, we have

$$\frac{SP}{SQ}+\frac{2}{l}\ SP-1. \qquad ...(5)$$

Similarly for the focal chord PS' R, we have

$$\frac{S'P}{S'R}=\frac{2}{l}\ S'P-1$$

Adding (5) and (6), we get

$$\frac{SP}{SQ}+\frac{S'P}{S'R}=\frac{2}{l}\ (SP + S'P) - 2.$$

But in the ellipse, SP + S'P = the sum of the focal distances of the point P = the length of the major axis = 2a, say.

$\therefore\ \dfrac{SP}{SQ}+\dfrac{S'P}{S'R}=\dfrac{2.2a}{l}-2$ which is a constant and so is independent of the position of P.

Example 11:

Prove that the perpendicular focal chords of a rectangular hyperbola are equal.

Solution :

Let PSP' and QSQ' be two perpendicular focal chords. Then proceeding as in Ex. 2 part (iii), we have

$$PP'=\frac{2l}{1-e^2\cos^2\alpha'}\quad QQ'=\frac{2l}{1-e^2\sin^2\alpha}.$$

For rectangular hyperbola, $e = \sqrt{2}$.

$$\therefore \quad PP' = \frac{2l}{1-2\cos^2\alpha} = \frac{2l}{\cos 2\alpha} = \frac{2l}{\cos 2\alpha} \qquad \text{(in magnitude)},$$

$$\text{and} \quad QQ' = \frac{2l}{1-2\sin^2\alpha} = \frac{2l}{\cos 2\alpha}.$$

$$\therefore \quad PP' = QQ'.$$

Example 12:

If the circle $r + 2a\cos\theta = 0$ cuts the conic $l/r = 1 + e\cos(\theta - \alpha)$ in four points, find the equation in r which determines the distances of these four points from the pole. Show that if the algebraic sum of these four distances is equal to 2a, the centricity is equal to 2 cos α.

Solution :

The equation of the conic given by

$$l/r = 1 + e\cos(\theta - \alpha)$$

$$\text{or} \quad l/r = 1 + e\cos\alpha\cos\theta + e\sin\theta\sin\alpha \qquad \text{...(1)}$$

and the equation of the circle is given by

$$r = -2a\cos\theta. \qquad \text{...(2)}$$

The radii-vectors of the points of intersection of (1) and (2) are given by

$$\left\{\frac{l}{r} - 1 - e\cos\alpha\left(-\frac{1}{2a}\right)\right\} = e\sin\alpha\left[\sqrt{\left(1-\frac{r^2}{4a^2}\right)}\right]$$

$$\Rightarrow \quad [2al - 2ar + er^2\cos\alpha]^2 = 4r^2a^2e^2\sin^2\alpha\left(\frac{4a^2-r^2}{4a^2}\right)$$

$$\Rightarrow \quad r^4(e^2\cos^2\alpha + e^2\sin^2\alpha) - 4ae\cos\alpha\, r^3$$
$$+ r^2(4a^2 - 4e^2\sin^2\alpha + 4ael\cos\alpha) - 8a^2lr + 4a^2l^2 = 0$$

$$\Rightarrow \quad e^2r^4 - 4ae\cos\alpha . r^3 + r^2(4a^2 - 4e^2\sin^2\alpha + 4ael\cos\alpha)$$
$$- 8a^2lr + 4a^2l^2 = 0 \qquad \text{...(3)}$$

The equation (3) gives the distances of the four points of intersection of (1) and (2) from the pole. Let these distances be r_1, r_2, r_3 and r_4.

$$\text{Then we have } r_1 + r_2 + r_3 + r_4 = \frac{4ae\cos\alpha}{e^2} - \frac{4a\cos\alpha}{e}.$$

Now by question, the algebraic sum of these four focal distances r_1, r_2, r_3 and r_4 is 2a.

Example 13:

A straight line drawn through the common focus S of a number of conics meets them in the points P_1, P_2, P_3,... On it is taken a point Q such that the reciprocal of SQ is equal to the sum of the reciprocals of SP_1, SP_2, SP_3,... Prove that the locus of Q is a conic section whose focus is S and the reciprocal of whose latus rectum is equal to the sum of the reciprocals of the latera recta of the given conics.

Solution:

Taking the common focus S as the pole and the common axis as the initial line the equations to the conics are

$$l_n/r = 1 + e_n \cos\theta \qquad \text{(where n = 1, 2, 3,...)}. \qquad ...(1)$$

Let a straight line drawn S make angle β with the common axis of the conics.

Suppose this straight line meets the $l_n/r = 1 + e_n \cos\theta$ at the point P_n, where the suffix n takes the values 1, 2, 3,.... Since the points P_n (n = 1, 2,...) lie on the same straight line, therefore their vectorial angles are the same. Let (r_n, β) the coordinates of the point P_n which lies on the conic $l_n/r = 1 + e_n \cos\theta$.

Then $l_n/r_n = 1 + e_n \cos\beta$, (r = 1, 2, ...). ...(2)

Suppose Q is the point (R, β) on this line. Then according to the question

$$\frac{1}{SQ} = \Sigma\frac{1}{SP_n}, \text{ or } \frac{1}{R} = \Sigma\frac{1}{r_n}$$

$$\Rightarrow \quad \frac{1}{R} = \Sigma\frac{1 + e_n \cos\beta}{l_n}, \text{ (n = 1, 2,...)}$$

$$= \frac{1 + e_1 \cos\beta}{l_1} + \frac{1 + e_2 \cos\beta}{l_2}$$

$$= \left(\frac{1}{l_1} + \frac{1}{l_2} + ...\right) + \left(\frac{e_1}{l_1} + \frac{e_2}{l_2} + ...\right)\cos\beta$$

$$= \frac{1}{L} + \frac{1}{K}\cos\beta \text{ where } \frac{1}{L} = \frac{1}{l_1} + \frac{1}{l_2} + ...$$

$$\Rightarrow \quad \frac{L}{R} = 1 + E\cos\beta, \text{ where } E = \frac{L}{K}.$$

$\therefore$ the locus of Q (R, β) is $\frac{L}{r} = 1 + E \cos \theta.$ ...(3)

This is the equation of a conic with focus S, semi latus rectum L and respectively E.

Again $\frac{1}{L} = \frac{1}{l_1} + \frac{1}{l_2} + ...$

$\Rightarrow \frac{1}{2L} = \frac{1}{2l_1} + \frac{1}{2l_2} + ...$

i.e., the reciprocal of the latus rectum of the conic (3) is equal to the sum of the reciprocals of the latera recta of the given conics (1). This proves the required result.

Example 14(a):

A point P moves, so that the sum of its distances from two fixed points S, S' is constant and equal to 2a. Show that P lies on the conic $\frac{a\left(1-e^2\right)}{r} = 1$ $- e \cos \theta$ *referred to S as pole and SS' as initial line, SS' being equal to 2ae.*

Solution :

Let us take S as the pole and SS' as the initial line, let the polar coordinates of P be (r, θ).

$\therefore$ PS = r, and PS'

= 2a – r because PS + PS'

= 2a (given).

Also SS' = 2ae (given).

In Δ PSS', we have

$$\cos \theta = \frac{r^2 + (2ae)^2 - (2a - r)^2}{2.r.(2ae)} \quad \text{[by cosine formula]}$$

$$= \frac{r + a\,(e^2 - 1)}{er}$$

$$\therefore \quad e \cos \theta = 1 + \frac{a\,(e^2 - 1)}{r},$$

or $\frac{a\,(e^2 - 1)}{r} = 1 - e \cos \theta$

which is the required locus of P and is a conic.

Example 14(b):

A circle passing through the focus of a conic whose latus rectum is 2l meets the conic in four points whose distances from the focus are r_1, r_2, r_3 and r_4, prove that

$$\frac{1}{r_1}+\frac{1}{r_2}+\frac{1}{r_3}+\frac{1}{r_4}=\frac{2}{l}$$

Solution :

The four values of r given by the equation (3) are the total focal distances of the four points intersection of the circle and the conic.

Let the four values of r given by (3) be r_1, r_2, r_3 and r_4.

Then,

$$r_2r_3r_4 + r_1r_3r_4 + r_1r_2r_4 + r_1r_2r_2 = -\frac{\text{the coeff. of r in (3)}}{\text{the coeff. of } r^4 \text{ in (3)}} \quad \text{...(4)}$$

$$\text{and } r_1r_2r_3r_4 = \frac{\text{constant term}}{\text{coeff. of } r^4} . \quad \text{...(5)}$$

Dividing (4) by (5), we get

$$\frac{1}{r_1}+\frac{1}{r_2}+\frac{1}{r_3}+\frac{1}{r_4} = -\frac{\text{ceoff. of r}}{\text{constant term}} = \frac{8a^2l}{4a^2l^2} = \frac{2}{l} .$$

Example 14(c):

(a) Prove that the condition that the line

$$l/r = A \cos\theta + B \sin\theta$$

may touch the conic $l/r = 1 + e\cos(\theta - \alpha)$ is

$$A^2 + B^2 - 2e(A\cos\alpha + B\sin\alpha) + e^2 - 1 = 0.$$

Solution :

Let the given line

$$l/r = A\cos\theta + B\sin\theta \quad \text{...(1)}$$

is a tangent to the given conic at the point whose vectorial angle is β. Then (1) should be identical with the tangent to the given conic at the point β whose equation is as

$$l/r = \cos(\theta - \beta) + e\cos(\theta - \alpha)$$

$\Rightarrow$ $l/r = (\cos\beta + e\cos\alpha)\cos\theta + (\sin\beta + e\sin\alpha)\sin\theta$ ...(2)

Comparing (1) and (2), we have

$$1 = \frac{\cos\beta + e\cos\alpha}{A} = \frac{\sin\beta + e\sin\alpha}{B}$$

or $A - e\cos\alpha = \cos\beta$, and $B - e\sin\alpha = \sin\beta$.

Squaring and adding, we have

$$(A - e\cos\alpha)^2 + (B - e\sin\alpha)^2 = 1$$

or $A^2 + B^2 - 2e(A\cos\alpha + B\sin\alpha) + e^2 - 1 = 0$,

as the required condition.

Example 15(a):

If POP' be a chord of a conic through a fixed point O, prove that $\tan\frac{1}{2}P'SO \tan\frac{1}{2}PSO$ *is constant, S being a focus of the conic.*

Solution :

Let the conic be $l/r = 1 + e\cos\theta$, the focus S being at the pole. Let O be the fixed point (a, α). Suppose the vectorial angles of the ends P and P' of any chord POP' passing through the fixed point O are β and α respectively. Then $\angle P'SO = \gamma - \alpha$ and $\angle PSO = \alpha - \beta$.

The equation of the chord PP' is

$$l/r = e\cos\theta + \sec\frac{1}{2}(\beta - \gamma)\cos\left\{\theta - \frac{1}{2}(\beta + \gamma)\right\}.$$

Since it passes through the point O (a, α), therefore

$$l/a = e\cos\alpha + \sec\frac{1}{2}(\beta - \gamma)\cos\left\{\alpha - \frac{1}{2}(\beta + \gamma)\right\}$$

$$\Rightarrow \frac{\cos\left\{\alpha - \frac{1}{2}(\beta + \gamma)\right\}}{\cos\frac{1}{2}(\beta - \gamma)} = \frac{1}{a} - a\cos\alpha$$

= k, say, where k is a constant because l, a, e and α are all constants

$$\Rightarrow \frac{\cos\left\{\frac{1}{2}(\alpha - \gamma) + \frac{1}{2}(\alpha - \beta)\right\}}{\cos\left\{\frac{1}{2}(\alpha - \gamma) - \frac{1}{2}(\alpha - \beta)\right\}} = k$$

$$\Rightarrow \frac{\cos\frac{1}{2}(\alpha-\gamma)\cos\frac{1}{2}(\alpha-\beta)-\sin\frac{1}{2}(\alpha-\gamma)\sin\frac{1}{2}(\alpha-\beta)}{\cos\frac{1}{2}(\alpha-\lambda)\cos\frac{1}{2}(\alpha-\beta)+\sin\frac{1}{2}(\alpha-\gamma)\sin\frac{1}{2}(\alpha-\beta)} = k$$

$$\Rightarrow \frac{1-\tan\frac{1}{2}(\alpha-\gamma)\tan\frac{1}{2}(\alpha-\beta)}{1+\tan\frac{1}{2}(\alpha-\gamma)\tan\frac{1}{2}(\alpha-\beta)} = \frac{k}{1}$$

Applying componendo and dividendo, we have

$$\frac{2\tan\frac{1}{2}(\alpha-\gamma)\tan\frac{1}{2}(\alpha-\beta)}{2} = \frac{k-1}{k+1}$$

$$\Rightarrow \tan\frac{1}{2}(\gamma-\alpha)\tan\frac{1}{2}(\alpha-\beta) = \frac{1-k}{1+k}$$

$$\Rightarrow \tan\frac{1}{2}\text{P'SO}\tan\frac{1}{2}\text{PSO} = \text{constant.}$$

Example 15(b):

Prove that the line $\frac{l}{r} = \cos(\theta-\alpha) + e\cos(\theta-\gamma)$ *is the tangent to the conic* $\frac{l}{r} = 1 + e\cos(\theta-\gamma)$ *at the point for which* $\theta = \alpha$.

Solution :

The line interests the conic where

$$\cos(\theta-\alpha) = 1$$

$$\Rightarrow \theta-\alpha = 2n\pi, \text{ where n is any integer}$$

$$\Rightarrow \theta = \alpha + 2n\pi.$$

For every integral value of n, $\theta = \alpha + 2n\pi$ represents the same point on the conic whose vectorial angle is α. Thus the given line meets the conic only at the point α and hence is a tangent to the conic at the point for which $\theta = \alpha$.

Example 15(c):

Show that the two conics

$$l_1/r = 1 + e_1\cos\theta$$

and $l_2/r = 1 + e_2\cos(\theta-\alpha)$

will touch one another if

$$l_1^2\left(1-e_2^2\right)+l_2^2\left(1-e_1^2\right)=2l_1l_2\left(1-e_1e_2\cos\alpha\right).$$

Solution :

Suppose the given conic touch one another at the point whose vectorial angle is β. The equations of the tangents at the common point 'β' to the two conics are

$$l_1/r = \cos(\theta - \beta) + e_1 \cos\theta, \qquad ...(1)$$

and $l_2/r = \cos(\theta - \beta) + e_2 \cos(\theta - \alpha)$. ...(2)

The equations (1) and (2) may be written as

$$l_1/r = (\cos\beta + e_2 \cos\alpha)\cos\theta$$

and $l_2/r = (\cos\beta + e_2\cos\alpha)\cos\theta$ ...(3)

$$+ (\sin\beta + e_2 \sin\alpha)\sin\theta. \qquad ...(4)$$

The equations (3) and (4) should be identical. Hence comparing them, we have

$$\frac{l_1}{l_2}=\frac{\cos\beta+e_1}{\cos\beta+e_2\cos\alpha}=\frac{\sin\beta}{(\sin\beta+e_2\sin\alpha)}$$

$\Rightarrow$ $(l_1 - l_2)\cos\beta = -l_1e_2\cos\alpha + l_2e_1,$

and $(l_1 - l_2)\sin$

$$\beta = l_1e_2 \sin\alpha.$$

Squaring and adding, we get

$$(l_1 - l_2)^2 = l_1^2 e_2^2 + l_2^2 e_1^2 - 2l_1l_2 e_1e_2 \cos\alpha,$$

$$\Rightarrow l_1^2 (1-e_2^2)+l_2^2(1-e_1^2)=2l_1l_2(1-e_1e_2\cos\alpha),$$

as the required condition.

Example 15(d):

A circle of given radius passing through the focus S of a given conic interests it in A, B, C, D; show that SA.SB.SC.CD is constant.

Solution :

Referred to the focus S as the pole and the axis as the initial line let the equation of the conic be given as

$$l/r = 1 + e\cos\theta. \qquad ...(1)$$

The equation of any circle of given radius a and passing through the pole (*i.e.*, the focus) is given as

$$r = 2a \cos (\theta - \theta_1)$$

$$\Rightarrow r = 2a \cos \theta \cos \theta_1 + 2a \sin \theta \sin \theta_1. \quad ...(2)$$

The focal distances of the points of intersection where these meet are given by eliminating θ between (1) and (2). From (1),

$$\cos \theta = (l - r)/(re).$$

Putting it in (2), we get

$$r = 2a \cos \theta_1 \frac{l-r}{re} + 2a \sin \theta_1 \sqrt{\left\{1 - \left(\frac{l-r}{re}\right)^2\right\}}$$

$$\Rightarrow \left\{r - 2a\left(\frac{l-r}{re}\right)\cos \theta_1\right\}^2 = 4a^2 \sin^2 \theta_1 \left\{1 - \frac{(l-r)^2}{r^2e^2}\right\}$$

$$\Rightarrow e^2r^4 + 4ear^3 \cos \theta_1 + 4ar^2 (a - el \cos \theta_1 - e^2a \sin^2 \theta_1)$$

$$- 8a^2lr + 4a^2l^2 = 0. \quad ...(3)$$

The equation (3) is of degree four in r and so it gives four values of r. These four values of r are SA, SB, SC and SD *i.e.*, the radii vectors of the four points of intersection of (1) and (2).

$$\therefore SA.SB.SC.SD = \frac{\text{the costant term in (3)}}{\text{the coefficient of } r^4 \text{ in (3)}} = \frac{4a^2l^2}{e^2}$$

which is constant *i.e.*, independent of θ_1.

Example 16:

Show that the locus of the point of intersection of two tangents of the parabola l/r = 1 cos θ, which cut one another at a constant angle α is the hyperbola, l/r = cos α + cos θ.

Solution :

The given parabola is

$$l/r = 1 + \cos \theta. \quad ...(1)$$

Consider two points β and γ on (1). The tangents at these points are

$$l/r = \cos (\theta - \beta) + \cos \theta \quad ...(2)$$

or $\quad l = (\cos \beta + \gamma)\, x + y \sin \beta \quad ...(2')$

and $l/r = \cos(\theta - 1) + \cos\theta$ ...(3)

$\Rightarrow \qquad l = (\cos\gamma + 1)\,x + y\sin\gamma.$ ...(3')

$$\therefore \quad m_1 = \text{the slope of } (2') = -\frac{\cos\beta + 1}{\sin\beta}$$

$$= -\frac{2\cos^2\frac{1}{2}\beta}{2\sin\frac{1}{2}\beta\cos\frac{1}{2}\beta} = -\cot\frac{1}{2}\beta$$

and m_2 = the slope (3')

$$= -\frac{\cos\gamma + 1}{\sin\gamma} = -\cot\frac{1}{2}\gamma.$$

It is given that the angle between the tangents (2) and (3) is a. Therefore using the formula

$$\tan\alpha = \frac{m_1 - m_2}{1 + m_1 m_2}, \text{ we have}$$

$$\tan\alpha = \frac{-\cot\frac{1}{2}\beta + \cot\frac{1}{2}\gamma}{1 + \cot\frac{1}{2}\beta\cot\frac{1}{2}\gamma} = \tan\left(\frac{\beta - \gamma}{2}\right).$$

$\therefore \qquad \alpha = (\beta - \gamma)/2.$ ...(4)

Now to find the locus of the points of intersection of the tangents (2) and (3), we have to eliminate β and γ between (2), (3) and (4). Equating the values of l/r from (2) and (3), we have

$$\cos(\theta - \beta) + \cos\theta = \cos(\theta - \gamma) + \cos\theta,$$

$\Rightarrow \cos(\theta - \beta) = \cos(\theta - \gamma)$ or $\theta - \beta = -(\theta - \gamma)$

[$\therefore$ if $\theta - \beta = \theta - \gamma$, then $\beta = \gamma$ which is not so]

$\Rightarrow 2\theta = \beta + \gamma$ or $\theta = \frac{1}{2}(\beta + \gamma).$ ...(5)

Adding (4) and (5), we get $\theta + \alpha = \beta$.

Putting $\beta = \theta + \alpha$ in (2), we get

$l/r = \cos\alpha + \cos\theta,$ ...(6)

which is the required locus of the point of intersection of the tangents (2) and (3).

The equation (6) may be written as $(l\sec\alpha)/r = 1 + \sec\alpha\cos\theta$ which is the equation of a conic whose eccentricity is sec α. Since sec $\alpha > 1$, therefore the conic is a hyperbola.

Example 17:

PS'P is a focal chord of a conic Prove that the locus of its middle point is a conic of the same kind as the original conic.

Solution :

Let the equation of the conic given as

$$l/r = 1 + e\cos\theta. \qquad ...(1)$$

Now net the vectorial angle of P be α, then that of P' is $\pi + \alpha$.

Since P, P' lie on (1),

$$\therefore\ 1/SP = 1 + e\cos\alpha,$$

and $1/SP' = 1 - e\cos\alpha$.

Let M is the mid-point of PP'.

If the polar coordinations of M are (r_1, α), then

$$r_1 = SM = SP - MP = SP - \frac{1}{2}(SP + SP')$$

$$= \frac{1}{2}(SP - SP') = \frac{1}{2}\left(\frac{l}{1+e\cos\alpha} - \frac{l}{1-e\cos\alpha}\right)$$

$$= \frac{1 - le\cos\alpha}{1 - e^2\cos^2\alpha}.$$

Generalizing *i.e.*, replacing r_1 by r and α by θ, the locus of the joint M (r_1, α) is

$$r(1 - e^2\cos^2\theta) = -le\cos\theta. \qquad ...(2)$$

Example 18(a):

A chord PQ of a conic whose eccentricity is e and semi latus rectum l subtends a right angle at the focus S, show that

$$\left(\frac{1}{SP} - \frac{1}{l}\right)^2 + \left(\frac{1}{SQ} - \frac{1}{l}\right)^2 = \frac{e^2}{l^2}.$$

Solution :

Le the equation of the conic be

$$l/r = 1 + e\cos\theta. \qquad ...(1)$$

It is given that the chord PQ subtends a right angle at the focus S.

Therefore if the vectorial angle of P is α, then $\angle Q = \frac{\pi}{2} + \alpha$

$$\therefore \frac{1}{SP} + 1 + e \cos \alpha$$

and $$\frac{l}{SQ} = 1 + e \cos \left(\frac{1}{2}\pi + \alpha\right)$$

$$= 1 - e \sin \alpha.$$

$$\therefore \left(\frac{1}{SP} - \frac{1}{l}\right)^2 + \left(\frac{1}{SQ} - \frac{1}{l}\right)^2$$

$$= \left(\frac{1 + e \cos \alpha}{l} - \frac{1}{l}\right)^2 + \left(\frac{1 - e \sin \alpha}{l} - \frac{1}{l}\right)^2$$

$$= \left(\frac{e \cos \alpha}{1}\right) + \left(\frac{- e \sin^2 \alpha}{1}\right)$$

$$= \frac{e^2}{l^2}\left(\cos^2 \alpha + \sin^2 \alpha\right) = \frac{e^2}{l^2}.$$ Hence

Example 18(b):

If PSP' and QSQ' be two perpendicular focal chords of a conic, prove that $\frac{1}{SP.SP'} + \frac{1}{QS.QS''}$ *is constant.*

Solution :

We have for the focal chord PSP',

$$\frac{1}{SP.SP'} = \frac{1 + e \cos \alpha}{l} \cdot \frac{1 - e \cos \alpha}{l} = \frac{1 - e^2 \cos^2 \alpha}{l^2}. \qquad ...(1)$$

Replacing SP by SQ, SP' by SQ' and α by $\frac{1}{2}\pi + \alpha$ in (1), we have for the focal chord QSQ',

$$\frac{1}{SQ.SQ'} = \frac{1 - e^2 \cos^2 \left(\frac{1}{2}\pi + \alpha\right)}{l^2} = \frac{1 - e^2 \sin^2 \alpha}{l^2} \qquad ...(2)$$

Adding (1) and (2), we have

$$\frac{1}{SP.SP'} + \frac{1}{SQ.SQ'} = \frac{2 - c^2}{l^2}$$ which is constant.

Example 18(c):

A circle of given radius passing through the focus S of a given conic interests it in A, B, C, D; show that SA.SB.SC.CD is constant

Solution :

Referred to the focus S as the pole and the axis as the initial line let the equation of the conic be given as

$$l/r = 1 + e \cos \theta. \qquad ...(1)$$

The equation of any circle of given radius a and passing through the pole (*i.e.*, the focus) is given as

$$r = 2a \cos (\theta - \theta_1)$$

$$\Rightarrow \ r = 2a \cos \theta \cos \theta_1 + 2a \sin \theta \sin \theta_1. \qquad ...(2)$$

The focal distances of the points of intersection where these meet are given by eliminating θ between (1) and (2). From (1),

$$\cos \theta = (l - r)/(re).$$

Putting it in (2), we get

$$r = 2a \cos \theta_1 \frac{l-r}{re} + 2a \sin \theta_1 \sqrt{\left\{1 - \left(\frac{1-r}{re}\right)^2\right\}}$$

$$\Rightarrow \ \left\{r - 2a\left(\frac{l-r}{re}\right)\cos \theta_1\right\}^2 = 4a^2 \sin^2 \theta_1 \left\{1 - \frac{(l-r)^2}{r^2e^2}\right\}$$

$$\Rightarrow \ e^2r^4 + 4ear^3 \cos \theta_1 + 4ar^2 (a - el \cos \theta_1 - e^2a \sin^2 \theta_1)$$
$$- 8a^2lr + 4a^2l^2 = 0. \qquad ...(3)$$

The equation (3) is of degree four in r and so it gives four values of r. These four values of r are SA, SB, SC and SD *i.e.*, the radii vectors of the four points of intersection of (1) and (2).

$$\therefore \ SA.SB.SC.SD = \frac{\text{the costant term in (3)}}{\text{the coefficient of } r^4 \text{ in (3)}} = \frac{4a^2l^2}{e^2}$$

which is constant *i.e.*, independent of θ_1.

Example 19(a):

If PQ is chord of contact of tangenst drawn from a point T to a conic whose focus is S, prove that

(1) $ST^2 = SP.SQ$, if the conic is a parabola;

(2) $\dfrac{1}{SP.SQ} - \dfrac{1}{ST^2} = \dfrac{1}{b^2}\sin^2\dfrac{PSQ}{2}$ if the conic is central conic and b is its semi-minor axis.

Solution:

(1) Let the equation of a parabola be

$$l/r = 1 + \cos\theta, \qquad ...(1)$$

the focus S being at the pole.

Let a, b thevectorial angles of P, Q respectively. Then

$$\left.\begin{aligned} l/SP &= 1 + \cos\alpha = 2\cos^2\frac{1}{2}\alpha \text{ i.e., } SP = (l/2)\sec^2\frac{1}{2}\alpha, \\ l/SQ &= 1 + \cos\beta = 2\cos^2\frac{1}{2}\beta \text{ i.e., } SQ = (l/2)\sec^2\frac{1}{2}\beta. \end{aligned}\right\} \qquad ...(2)$$

Now let T (r', θ') be the pointof intersection of the tangents at P and Q. Then

$$ST = (l/2)\sec\frac{1}{2}\alpha\sec\frac{1}{2}\beta.$$

$$\therefore\ ST^2 = (l^2/4)\sec^2\frac{1}{2}\alpha\sec^2\frac{1}{2}\beta = SP.\ SQ, \text{ using (2).}$$

(3) Let the equation of a central conic be

$$l/r = 1 + e\cos\theta, \qquad ...(1)$$

the focus S being at the pole.

Let α, β be the vectorial angles of P, Q respectively.

Since P and Q lie on (1), we have

$$l/SP = 1 + e\cos\alpha,\quad l/SQ = 1 + e\cos\beta.$$

Multiplying,

$$\frac{l^2}{SP.SQ} = 1 + e(\cos\alpha + \cos\beta) + e^2\cos\alpha\cos\beta$$

$$= 1 + 2e\cos\frac{1}{2}(\alpha+\beta)\cos\frac{1}{2}(\alpha-\beta) + e^2\cos\alpha\cos\beta. \qquad ...(2)$$

Also clearly $\angle PSQ = \alpha - \beta$. ...(3)

Now suppose T (r', θ') is the point of intersection of the tangents at P and Q. Then proceeding as in § 9, we have

$$l/r' = l/ST = \cos\frac{1}{2}(\alpha-\beta) + e\cos\frac{1}{2}(\alpha+\beta). \qquad ...(4)$$

Now from (2) and (4), we have

$$= 1 + 2e\cos\frac{1}{2}(\alpha+\beta)\cos\frac{1}{2}(\alpha-\beta) + e^2\cos a\cos b$$

$$-\cos^2\frac{1}{2}(\alpha-\beta) - e^2\cos^2\frac{1}{2}(\alpha+\beta) - 2e\cos\frac{1}{2}(\alpha+\beta)\cos\frac{1}{2}(\alpha-\beta)$$

$$= \left\{1-\cos^2\frac{1}{2}(\alpha-\beta)\right\} - \frac{1}{2}e^2\left\{2\cos^2\frac{1}{2}(\alpha+\beta) - 2\cos\alpha\cos\beta\right\}$$

$$= \sin^2\frac{1}{2}(\alpha-\beta) - \frac{1}{2}e^2\{1+\cos(\alpha+\beta) - 2\cos\alpha\cos\beta\}$$

$$= \sin^2\frac{1}{2}(\alpha-\beta) - \frac{1}{2}e^2\{1+\cos\alpha\cos\beta - \sin\alpha\sin\beta - 2\cos\alpha\cos\beta\}$$

$$= \sin^2\frac{1}{2}(\alpha-\beta) - \frac{1}{2}e^2\{1-(\cos\alpha\cos\beta+\sin\alpha\sin\beta)\}$$

$$= \sin^2\frac{1}{2}(\alpha-\beta) - \frac{1}{2}e^2\{1-\cos(\alpha-\beta)\}$$

$$= \sin^2\frac{1}{2}(\alpha-\beta) - \frac{1}{2}e^2 . 2\sin^2\frac{1}{2}(\alpha-\beta)$$

$$= (1-e^2)\sin^2\frac{1}{2}(\alpha-\beta).$$

$$\therefore \quad \frac{1}{SP.SQ} - \frac{1}{ST^2} = \frac{\left(1-e^2\right)}{l^2}\sin^2\frac{1}{2}(\alpha-\beta). \qquad ...(5)$$

Now in a central conic, we have

$$l = \text{semi-latus rectum} = b^2/a, \text{ and } b^2 = a^2(1-e^2).$$

$$\therefore \quad \frac{1-e^2}{l^2} = \frac{1-e^2}{b^4/a^2} = \frac{a^2\left(1-e^2\right)}{b^4} = \frac{b^2}{b^4} = \frac{1}{b^2} \qquad ...(6)$$

Making use of the results (3) and (6) in (5), we have

$$\frac{1}{SP.SQ} - \frac{1}{ST^2} = \frac{1}{b^2} \sin^2 \frac{1}{2} PSQ.$$

Example 19(b):

Find the locus of the point of intersection of the tangents to the conic $l/r = 1 + e \cos \theta$ at points P and Q subject to the condition

$$1/SP + 1/SQ = 2/k,$$

k being a constant.

Solution:

Let α, β be the vectorial angles of P, Q respectvely.

Proceeding as in § 9, we have

$$\theta' = \frac{1}{2} (\alpha + \beta), \quad \text{...(1)}$$

and $$l/r' = \cos \frac{1}{2} (\alpha - \beta) + e \cos \frac{1}{2} (\alpha + \beta), \quad \text{...(2)}$$

where (r', θ') are the coordinates of the point of intersection T of the tangents at P Q.

Also $$l/SP = 1 + e \cos \alpha, \quad \text{...(3)}$$

and $$l/SQ = 1 + e \cos \beta. \quad \text{...(4)}$$

Given that $\frac{1}{SP} + \frac{1}{SQ} = \frac{2}{k}$. ...(5)

Substituting for 1/SP and 1/SQ from (3) and (4) in (5), we have

$$\frac{1}{l} [1 + e\cos \alpha + 1 + e \cos \beta] = \frac{2}{k}$$

$$\Rightarrow \quad e (\cos \alpha + \cos \beta) = (2l/k) - 2$$

$$\Rightarrow \quad 2e \cos \frac{1}{2} (\alpha + \beta) \cos \frac{1}{2} (\alpha - \beta) = 2 (l/k - 1)$$

$$\Rightarrow \quad e \cos \frac{1}{2} (\alpha + \beta) \{l/r' - e \cos \frac{1}{2} (\alpha + \beta)\} = l/k - 1$$

[putting for $\cos \frac{1}{2} (\alpha - \beta)$ from (2)]

$$\Rightarrow \quad e \cos \theta' \{1/r' - e \cos \theta'\} = l/k - 1.$$

[∵ from (1), $\theta' = \frac{1}{2} (\alpha + \beta)$]

∴ the locus of T(r', θ') is

$e \cos\theta \{l/r - e\cos\theta'\} = l/k - 1$.Example 44:

If PQ is chord of contact of tangenst drawn from a point T to a conic whose focus is S, prove that

(1) $ST^2 = SP.SQ$, if the conic is a parabola;

(2) $\frac{1}{SP.SQ} - \frac{1}{ST^2} = \frac{1}{b^2}\sin^2\frac{PSQ}{2}$ if the conic is central conic and b is its semi-minor axis.

Solution:

(1) Let the equation of a parabola be

$$l/r = 1 + \cos\theta, \qquad ...(1)$$

the focus S being at the pole.

Let a, b thevectorial angles of P, Q respectively. Then

$$\left.\begin{aligned} l/SP = 1+\cos\alpha = 2\cos^2\frac{1}{2}\alpha \text{ i.e., } SP = (l/2)\sec^2\frac{1}{2}\alpha, \\ l/SQ = 1+\cos\beta = 2\cos^2\frac{1}{2}\beta \text{ i.e., } SQ = (l/2)\sec^2\frac{1}{2}\beta. \end{aligned}\right\} \qquad ...(2)$$

Now let T (r', θ') be the pointof intersection of the tangents at P and Q. Then

$$ST = (l/2)\sec\frac{1}{2}\alpha\sec\frac{1}{2}\beta.$$

$$\therefore \quad ST^2 = (l^2/4)\sec^2\frac{1}{2}\alpha\sec^2\frac{1}{2}\beta = SP.\ SQ, \text{ using (2).}$$

(3) Let the equation of a central conic be

$$l/r = 1 + e\cos\theta, \qquad ...(1)$$

the focus S being at the pole.

Let α, β be the vectorial angles of P, Q respectively.

Since P and Q lie on (1), we have

$$l/SP = 1 + e\cos\alpha, \quad l/SQ = 1 + e\cos\beta.$$

Multiplying,

$$\frac{l^2}{SP.SQ} = 1 + e(\cos\alpha + \cos\beta) + e^2\cos\alpha\cos\beta$$

$$= 1 + 2e \cos \frac{1}{2} (\alpha + \beta) \cos \frac{1}{2} (\alpha - \beta) + e^2 \cos \alpha \cos \beta \quad ....(2)$$

Also clearly $\angle PSQ = \alpha - \beta$. ...(3)

Now suppose T (r', θ') is the point of intersection of the tangents at P and Q. Then proceeding as in § 9, we have

$$l/r' = l/ST = \cos \frac{1}{2} (\alpha - \beta) + e \cos \frac{1}{2} (\alpha + \beta). \quad ...(4)$$

Now from (2) and (4), we have

$$= 1 + 2e \cos \frac{1}{2} (\alpha + \beta) \cos \frac{1}{2} (\alpha - \beta) + e^2 \cos a \cos b$$

$$- \cos^2 \frac{1}{2} (\alpha - \beta) - e^2 \cos^2 \frac{1}{2} (\alpha + \beta) - 2e \cos \frac{1}{2} (\alpha + \beta) \cos \frac{1}{2} (\alpha - \beta)$$

$$= \left\{1 - \cos^2 \frac{1}{2} (\alpha - \beta)\right\} - \frac{1}{2} e^2 \left\{2 \cos^2 \frac{1}{2} (\alpha + \beta) - 2 \cos\alpha \cos\beta\right\}$$

$$= \sin^2 \frac{1}{2} (\alpha - \beta) - \frac{1}{2} e^2 \{1 + \cos (\alpha + \beta) - 2 \cos \alpha \cos \beta\}$$

$$= \sin^2 \frac{1}{2} (\alpha - \beta) - \frac{1}{2} e^2 \{1 + \cos \alpha \cos \beta - \sin \alpha \sin \beta - 2 \cos \alpha \cos \beta\}$$

$$= \sin^2 \frac{1}{2} (\alpha - \beta) - \frac{1}{2} e^2 \{1 - (\cos \alpha \cos \beta + \sin \alpha \sin \beta)\}$$

$$= \sin^2 \frac{1}{2} (\alpha - \beta) - \frac{1}{2} e^2 \{1 - \cos (\alpha - \beta)\}$$

$$= \sin^2 \frac{1}{2} (\alpha - \beta) - \frac{1}{2} e^2 . 2 \sin^2 \frac{1}{2} (\alpha - \beta)$$

$$= (1 - e^2) \sin^2 \frac{1}{2} (\alpha - \beta).$$

$$\therefore \quad \frac{1}{SP.SQ} - \frac{1}{ST^2} = \frac{(1 - e^2)}{l^2} \sin^2 \frac{1}{2}(\alpha - \beta). \quad ...(5)$$

Now in a central conic, we have

$$l = \text{semi-latus rectum} = b^2/a, \text{ and } b^2 = a^2 (1 - e^2).$$

$$\therefore \quad \frac{1-e^2}{l^2} = \frac{1-e^2}{b^4/a^2} = \frac{a^2\left(1-e^2\right)}{b^4} = \frac{b^2}{b^4} = \frac{1}{b^2} \qquad ...(6)$$

Making use of the results (3) and (6) in (5), we have

$$\frac{1}{SP.SQ} - \frac{1}{ST^2} = \frac{1}{b^2}\sin^2\frac{1}{2}PSQ.$$

Example 19(c):

Find the locus of the point of intersection of the tangents to the conic $l/r = 1 + e\cos\theta$ at points P and Q subject to the condition

$$1/SP + 1/SQ = 2/k,$$

k being a constant.

Solution:

Let α, β be the vectorial angles of P, Q respectvely.

Proceeding as in § 9, we have

$$\theta' = \frac{1}{2}(\alpha + \beta), \qquad ...(1)$$

and $$l/r' = \cos\frac{1}{2}(\alpha - \beta) + e\cos\frac{1}{2}(\alpha + \beta), \qquad ...(2)$$

where (r', θ') are the coordinates of the point of intersection T of the tangents at P Q.

Also $$l/SP = 1 + e\cos\alpha, \qquad ...(3)$$

and $$l/SQ = 1 + e\cos\beta. \qquad ...(4)$$

Given that $\frac{1}{SP} + \frac{1}{SQ} = \frac{2}{k}$. ...(5)

Substituting for 1/SP and 1/SQ from (3) and (4) in (5), we have

$$\frac{1}{l}[1 + e\cos\alpha + 1 + e\cos\beta] = \frac{2}{k}$$

$$\Rightarrow \quad e(\cos\alpha + \cos\beta) = (2l/k) - 2$$

$$\Rightarrow \quad 2e\cos\frac{1}{2}(\alpha+\beta)\cos\frac{1}{2}(\alpha-\beta) = 2(l/k - 1)$$

$$\Rightarrow \quad e\cos\frac{1}{2}(\alpha+\beta)\left\{l/r' - e\cos\frac{1}{2}(\alpha+\beta)\right\} = l/k - 1$$

[putting for $\cos\frac{1}{2}(\alpha - \beta)$ from (2)]

$\Rightarrow \qquad e \cos \theta' \{1/r' - e \cos \theta'\} = l/k - 1.$

$$\left[\because \quad \text{from (1), } \theta' = \frac{1}{2}(\alpha + \beta)\right]$$

$\therefore$ the locus of $T(r', \theta')$ is

$$e \cos \theta \{l/r - e \cos \theta'\} = l/k - 1.$$

Example 20:

Prove that the portion of the tangent intercepted between the conic and the directrix subtends a right angle at the corresponding focus.

Or

Let the tangent at any point P on a conic whose focus is S meet the directrix in K, show that the angle PSK is a right angle.

Solution :

Let the equation of the conic referred to the focus S as the pole be

$$l/r = 1 + e \cos \theta. \qquad ...(1)$$

The equation of the directrix corresponding to the focus S is

$$l/r = e \cos \theta. \qquad ...(2)$$

Let the vectorial angle of any point P on the conic be α. The tangent at P is

$$l/r = \cos(\theta - \alpha) + e \cos \theta. \qquad ...(3)$$

Now the vectorial angle θ of the point of intersection K of the tangent (3) and the directrix (2) is given by

$$\cos(\theta - \alpha) + e \cos \theta = e \cos \theta$$

[obtained on eliminating r between (2) and (3)]

or $\qquad \cos(\theta - \alpha) = 0.$

$\therefore \quad \theta \sim \alpha = 90^\circ$ *i.e.*, $\angle PSK = 90^\circ$.

Example 21:

PSP' is a focal chord of a conic; prove that the angle between the tangents at P and P' is $\tan^{-1}\left(\frac{2e \sin \alpha}{1 - e^2}\right)$ *where* α *is the angle between the chord and the major axis.*

Solution :

Referred to the focus S as the pole and the axis as the initial line, let the equation of the conic be

$$l/r = 1 + e \cos \theta. \qquad ...(1)$$

Since the focal chord PSP' is inclined at an angle α to the major axis, therefore if the vectorial angle of P is α then that of P' is $\pi + \alpha$.

The tangent at P is $l/r = \cos(\theta - \alpha) + e \cos\theta$

$\Rightarrow$ $l = r \cos\theta \cos\alpha + r \sin\alpha \sin a + er \cos\theta$

$\Rightarrow$ $l = x \cos\alpha + y \sin\alpha + ex$ [Changing to Cartesians]

$\Rightarrow$ $l = (\cos\alpha + e)\, x + y \sin\alpha.$

$\therefore$ m_1 = the slope of the tangent at P

$$= -\frac{\cos\alpha + e}{\sin\alpha}.$$

Similarly, m_2 = the slope of the tangent at P'

$$= -\frac{\cos(\alpha+\pi) + e}{\sin(\alpha+\pi)}$$

$$= -\frac{e - \cos\alpha}{-\sin\alpha} = \frac{e - \cos\alpha}{\sin\alpha}.$$

Now if β is the angle between the tangents at P and P', then

$$\tan\beta = \frac{m_1 \sim m_2}{1 + m_1 m_2} = \frac{(e - \cos\alpha)/\sin\alpha + (e + \cos\alpha)/\sin\alpha}{1 + \{(e - \cos\alpha)/\sin\alpha\}\,(-(e + \cos\alpha)/\sin\alpha\}}$$

$$= \frac{2e \sin\alpha}{\sin^2\alpha - (e^2 - \cos^2\alpha)} = \frac{2e \sin\alpha}{1 - e^2}.$$

$$\therefore \quad \beta = \tan^{-1}\left\{\frac{2e \sin\alpha}{1 - e^2}\right\}.$$

Example 22:

Prove that two points on the conic $l/r = 1 + e \cos\theta$ whose vectorial angles are α and β respectively will be the extremities of a diameter if

$$\frac{e+1}{e-1} = \tan\frac{\alpha}{2} \tan\frac{\beta}{2}.$$

Solution :

The given conic is

$$l/r = 1 + e \cos\theta. \quad ...(1)$$

Let the points P and Q be the extremities of a diameter of (1). Let the vectorial angles of P and Q be α and β respectively.

The tangent at P is

$$l/r = \cos(\theta - \alpha) + e\cos\theta$$

$$\Rightarrow\ l = (\cos\alpha + e)\,x + \sin\alpha\ y,$$

changing to cartesians.

$\therefore$ slope of the tangent at P

$$= -(\cos\alpha + e)/\sin\alpha. \qquad \text{...(2)}$$

Similarly the slope of the tangent at Q

$$= -(\cos\beta + e)/\sin\beta. \qquad \text{...(3)}$$

If PQ is a diameter, then the tangents at P and Q should be parallel and hence their slopes are equal *i.e.,*

$$-\frac{\cos\alpha + e}{\sin\alpha} = -\frac{\cos\beta + e}{\sin\beta}$$

$$\Rightarrow\ \cos\alpha\sin\beta + e\sin\beta = \sin\alpha\cos\beta + e\sin\alpha$$

$$\Rightarrow\ e(\sin\beta - \sin\alpha) = \sin(\alpha - \beta)$$

$$\Rightarrow\ 2e\cos\frac{1}{2}(\beta + \alpha)\sin\frac{1}{2}(\beta - \alpha) = 2\sin\frac{1}{2}(\alpha - \beta)\cos\frac{1}{2}(\alpha - \beta)$$

$$\Rightarrow\ e\cos\frac{1}{2}(\beta + \alpha) = -\cos\frac{1}{2}(\alpha - \beta)$$

$$\Rightarrow\ \frac{e}{1} = \frac{-\cos\frac{1}{2}(\alpha - \beta)}{\cos\frac{1}{2}(\beta + \alpha)}.$$

Applying componendo and dividendo, we have

$$\frac{e+1}{e-1} = \frac{-\cos\frac{1}{2}(\alpha-\beta) + \cos\frac{1}{2}(\beta+\alpha)}{-\cos\frac{1}{2}(\alpha-\beta) - \cos\frac{1}{2}(\beta+\alpha)} = \frac{\cos\frac{1}{2}(\alpha-\beta) - \cos\frac{1}{2}(\beta+\alpha)}{\cos\frac{1}{2}(\alpha-\beta) + \cos\frac{1}{2}(\beta+\alpha)}$$

$$= \frac{2\sin\frac{1}{2}\alpha\sin\frac{1}{2}\beta}{2\cos\frac{1}{2}\alpha\cos\frac{1}{2}\beta} = \tan\frac{1}{2}\alpha\tan\frac{1}{2}\beta.$$

Example 23(a):

A conic is described having the same focus and eccentricity as the conic $l/r = 1 + e\cos\theta$, and the two conics touch at the point $\theta = \alpha$; prove that *the length of its latus rectum is*

$$2l\,(1 - e^2)/(e^2 + 2e\cos\alpha + 1).$$

Solution :

The equation of the given conic is

$$l/r = 1 + e\cos\theta \qquad ...(1)$$

the focus being at the pole.

Let the equation of the conic having the same focus and eccentricity as the given conic be

$$l_1/r = 1 + e\cos(\theta - \gamma), \qquad ...(2)$$

where γ is angle of inclination of its axis to the initial line and l_1 is its semi-latus rectum. Since the conics (1) and (2) touch at the point $\theta = \alpha$, the tangents to them at the point α are the same lines.

The equation of the tangent to (1) at the points α is

$$l/r = \cos(\theta - \alpha) + \cos\theta$$

$$\Rightarrow \; l/r = (\cos\alpha + e)\cos\theta + \sin\alpha\sin\theta. \qquad ...(3)$$

The equation of the tangent to (2) at the point α is

$$l_1/r = \cos(\theta - \alpha) + e\cos(\theta - \gamma)$$

$$\Rightarrow \; l_1/r = (\cos\alpha + e\cos\gamma)\cos\theta + (\sin\alpha + e\sin\gamma)\sin\theta.$$
...(4)

Now the equations (3) and (4) are identical because they represent the same line. So comparing (3) and (4), we have

$$\frac{l}{l_1} = \frac{\cos\alpha + e}{\cos\alpha + e\cos\gamma} = \frac{\sin\alpha}{\sin\alpha + e\sin\gamma}$$

$$\therefore \; el\cos\gamma = (l_1 - l)\cos\alpha + el_1, \qquad ...(5)$$

and $$el\sin\gamma = (l_1 - l)\sin\alpha. \qquad ...(6)$$

To eliminate γ, squaring (5) and (6), and adding, we get

$$e^2l^2 = (l_1 - l)^2 + e^2\,l_1^2 + 2el_1\,(l_1 - l)\cos\alpha.$$

$$\Rightarrow \; -e^2\,(l_1^2 - l^2) = (l_1 - l)^2 + 2el_1\,(l_1 - l)\cos\alpha.$$

Since $l_1 \neq l$, therefore, dividing both sides by $l_1 - l$, we have

$$-e^2\,(l_1 + l) = (l_1 - l) + 2el_1\cos\alpha$$

$$\Rightarrow \; l_1\,(1 + 2e\cos\alpha + e^2) = l\,(1 - e^2).$$

∴ the length of the latus rectum of the conic (2)

$$= 2l_1 = \frac{2l\,(l - e^2)}{(1 + 2e \cos \alpha + e^2)}.$$

Example 23(b):

QR, a chord of the conic $l/r = 1 - e \cos \theta$, subtends a constant angle 2α at its focus S, sand SP, the bisector of the angle QSR, meets QR in P. Show that the locus of P is the conic

$$(l \cos \alpha)/r = 1 - e \cos \alpha \cos \theta.$$

Solution :

The given conic is $l/r = 1 - e \cos \theta$. ...(1)

the focus S being at the pole.

A chord QR of the conic (1) subtends a constant angle 2α at the focus S *i.e.*, $\Delta QSR = 2\alpha$. The bisector of the angle QSR meets QR in P; therefore

$$\Delta QSP = \angle PSR = \alpha.$$

Let the polar coordinates of the point P be (r_1, β), so that $SP = r_1$ and $\angle XSP = \beta$. The vectorial angles of the extremities Q and R of the chord QR are $\beta - \alpha$ and $\beta + \alpha$ respectively.

We know that the equation of the chord joining the points 'θ_1' and 'θ_2' of the cenic $l/r = 1 + e \cos \theta$ is

$$l/r = e \cos \theta + \sec \frac{1}{2} (\theta_1 - \theta_2) \cos \left\{\theta - \frac{1}{2}(\theta_1 + \theta_2)\right\} \quad ....(2)$$

Putting $\theta_1 = \beta - \alpha$ and $\theta_2 = \beta + \alpha$ in (2), we get the equation of the chord QR of the conic (1) as

$$l/r = -e \cos \theta + \sec \alpha \cos (\theta - \beta). \quad ...(3)$$

But the chord (3) passes throught the point P (r_1, β). Therefore

$$l/r_1 = -e \cos \beta + \sec \alpha \cos (\beta - \beta)$$

$$= -e \cos \beta + \sec \alpha. \quad ...(4)$$

Replacing r_1 by r and β by θ in the equation (4), the locus of the point P (r_1, β) is

$$l/r = -e \cos \theta + \sec \alpha$$

$$\Rightarrow \quad (l \cos \alpha)/r = 1 - e \cos \alpha \cos \theta,$$

which is a conic.

Example 23(c):

If the tangent at any point of an ellipse makes an angle α with its major axis and an angle β with the focal radius to the point of contact, show that $e \cos \alpha = \cos \beta$.

Solution :

Let the equation of an ellipse with focus S as pole be

$$l/r = 1 + e \cos \theta.$$

Suppose the vectorial angle of any point P on the ellipse $l/r = 1 + e \cos \theta$ is ϕ. The tangent at the point 'ϕ' is

$$l/r = \cos(\theta - \phi) + e \cos \theta$$

$$\Rightarrow \; l = (\cos \phi + e)\, x + \sin \phi y. \qquad ...(1)$$

since (1) makes an angle α with the major axis *i.e.*, with the initial line, therefore the slope of (1) is $\tan \alpha$.

$$\therefore \quad \tan \alpha = -\frac{\cos \phi + e}{\sin \phi}, \text{ or } \frac{\sin \alpha}{\cos \alpha} = -\frac{\cos \phi + e}{\sin \phi}$$

$$\Rightarrow \; \sin \alpha \sin \phi = -\cos \alpha \cos \phi - e \cos \alpha$$

$$\Rightarrow \; \cos(\alpha - \phi) = -e \cos \alpha$$

$$\Rightarrow \; \cos \beta = -e \cos \alpha \qquad ...(2)$$

(From the figure, $\phi + \beta = \alpha$)

If we consider the focal radius through the other focus *i.e.*, if we take $l/r = 1 - e \cos \theta$ as the equation of the ellipse, we shall get

$$\cos \beta = e \cos \alpha. \qquad ...(3)$$

In view of (2) and (3), we have in either case

$$e^2 \cos^2 \alpha = \cos^2 \beta.$$

Example 24:

Two conics have a common focus; prove that two of their common chords pass through the intersection of their directrices.

Solution :

Let the equations of the two conics having a common focus be

$$l/r = 1 + e_1 \cos \theta, \qquad ...(1)$$

$$\text{and } L/r = 1 + e_2 \cos(\theta - \alpha). \qquad ...(2)$$

The directrices of (1) and (2) respectively are

$l/r = e_1 \cos\theta$, $L/r = e_2 \cos(\theta - \alpha)$.

Now changing (1) to cartesians, we have

$$l = \sqrt{(x^2 + y^2)} + e_1 x$$

$$\Rightarrow (l - e_1 x)^2 = (x^2 + y^2),$$

making the equation rational.

Again on transforming to polars, we have

$$(l - e_1 r \cos\theta)^2 = r^2$$

$$\Rightarrow (l/r - e_1 \cos\theta)^2 - 1 = 0. \quad ...(3)$$

The equation (3) is thus the polar equation of the conic (1) put in the form which when transformed to cartesians gives rational cartesian equation of the conic.

Similarly the equation (2) can be written as

$$\{L/r - e_2 \cos(\theta - \alpha)\}^2 - 1 = 0. \quad ...(4)$$

Now any curve passing through the points of intersection of the two conics (3) and (4) is given by

$$\{(l/r - e_1 \cos\theta)^2 - 1\} + \lambda\,[\{L/r - e_2 \cos(\theta - \alpha)\}^2 - 1] = 0. \quad ...(5)$$

Clearly if $\lambda = -1$, (5) gives two lines, namely

$$l/r - e_1 \cos\theta = \pm\{L/r - e_2 \cos(\theta - \alpha)\},$$

which clearly pass through the point of intersection of the two directrices

$$l/r - e_1 \cos\theta = 0 \text{ and } L/r - e_2 \cos(\theta - \alpha) = 0.$$

Since the straight lines passing through the points of intersection of the conics (1) and (4) are their common chords, therefore the common chords of (1) and (2) pass through the intersection of their directrices.

Example 25:

If the tangent from a point P to the conic $l/r = 1 + e\cos\theta$ subtend the fixed angle p at the focus, prove that the locus of the middle point of SP is a conic of eccentricity e sec b.

Solution:

The conic is

$$l/r = 1 + e\cos\theta, \quad ...(1)$$

the focus S being at the pole.

Let PT be the tangent to the conic (1) from a point P. Let the polar coordinates of P be (r', α). If M be the middle point of SP, the polar coordinates of M are (r'/2, α).

Since the tangent PT subtends an angle β at the focus S *i.e.*, ∠PST = β, therefore the vectoria angle of the point of contact T is α + β.

Now the tangent to (1) at the point 'α + β' is

$$l/r = \cos(\theta - \alpha - \beta) + e \cos\theta. \qquad ...(2)$$

Since the straight line (2) passes through the point P (r', α), therefore $l/r' = \cos(\alpha - \alpha - \beta) + e \cos\alpha = \cos\beta + e\cos\alpha$

$$\Rightarrow \quad \frac{l/2}{r'/2} = \cos\beta + e\cos\alpha. \qquad ...(3)$$

To obtain the locus of the point M (r'/2, α), we replace r'/2 by r and α by θ in the equation (3). Therefore the locus of the point M is

$$\frac{l/2}{r} = \cos\beta + e\cos\theta = \cos\beta\,(1 + e\sec\beta\cos\theta)$$

$$\Rightarrow \quad \frac{(l\sec\beta)/2}{r} = 1 + e\sec\beta\cos\theta,$$

which is a conic of eccentricity e sec θ.

Example 26:

Show that the locus of the intersection of two perpendicular tangents one drawn to each of the two parabolas with a common focus whose axes are neither coincident nor perpendicular is a conic.

Solution:

Take the common focus of the two parabolas as the pole. Let one parabola be

$$l/r = 1 + \cos\theta, \qquad ...(1)$$

the axis of the parabola being taken as the initial line.

Suppose the axis of the other parabola in inclined at an angle φ to the initial line *i.e.*, the axis of (1). Then its equation is

$$l'/r = 1 + \cos(\theta + \phi). \qquad ...(2)$$

Note that $\phi \neq 0$ and $\phi \neq \pi/2$.

Suppose P is a point 'α' on (1) and Q is a point 'β' on (2).

The equation of the tangent to the parabola (1) at the point 'α' is

$$l/r = \cos(\theta - \alpha) + \cos\theta \qquad ...(3)$$

$$\Rightarrow \quad l = (1 + \cos\alpha)\, x + y \sin\alpha,$$

changing to cartesians.

Its slope $m = -\dfrac{1 + \cos\alpha}{\sin\alpha} = -\cos\dfrac{1}{2}\alpha.$

The equation of the tangent to the parabola (2) at the point 'β' is

$$l'/r = \cos(\theta - \beta) + \cos(\theta - \phi) \qquad ...(4)$$

$$\Rightarrow \quad l' = (\cos\beta + \cos\phi)\, x + (\sin\beta + \sin\phi)\, y,$$

changing to cartesians.

Its slope $m' = -\dfrac{\cos\beta + \cos\phi}{\sin\beta + \sin\phi}$

$$= -\frac{2\cos\frac{1}{2}(\beta + \phi)\cos\frac{1}{2}(\beta - \phi)}{2\sin\frac{1}{2}(\beta + \phi)\cos\frac{1}{2}(\beta - \phi)} = -\cot\frac{1}{2}(\beta + \phi).$$

According to the question, the tangents (3) and (4) are to be perpendicular.

$$\therefore \quad mm' = -1 \text{ i.e., } \cot\frac{1}{2}\alpha \cot\frac{1}{2}(\beta + \phi) = -1$$

$$\Rightarrow \quad \cos\frac{1}{2}\alpha\cos\frac{1}{2}(\beta + \phi) + \sin\frac{1}{2}\alpha\sin\frac{1}{2}(\beta + \phi) = 0$$

$$\Rightarrow \quad \cos\left\{\frac{1}{2}\alpha - \frac{1}{2}(\beta + \phi)\right\} = 0.$$

$$\therefore \quad \frac{1}{2}\alpha - \frac{1}{2}(\beta + \phi) = \pm\frac{1}{2}\pi, \quad \text{or } \alpha - \beta = \phi \pm \pi. \qquad ...(5)$$

It is required to find the locus of the point of intersection of (3) and (4). Hence we are to eliminate the variables α, β between (3), (4) and (5).

Adding (3) and (4), we get

$$(l + l')/r = \cos(\theta - \alpha) + \cos(\theta - \beta) + \cos\theta + \cos(\theta - \phi)$$

$$= 2\cos\left\{\theta - \frac{1}{2}(\alpha + \beta)\right\}\cos\frac{1}{2}(\alpha - \beta) + 2\cos\left(\theta - \frac{1}{2}\phi\right)\cos\frac{1}{2}\phi$$

$$= \pm 2\cos\left\{\theta - \frac{1}{2}(\alpha + \beta)\right\}\sin\frac{1}{2}\phi + 2\cos\left(\theta - \frac{1}{2}\phi\right)\cos\frac{1}{2}\phi \text{ [using (5)]}$$

Dividing by $\sin \frac{1}{2}\phi$ and transposing, we get

$$\frac{(l + l')}{r} \operatorname{cosec} \frac{1}{2}\phi - 2 \cos\left(\theta - \frac{1}{2}\phi\right) \cot \frac{1}{2}\phi$$

$$= \pm 2 \cos \left\{\theta - \frac{1}{2}(\alpha + \beta)\right\}. \qquad ...(6)$$

Subtracting (4) from (3), we have

$$(l + l')/r = \cos(\theta - \alpha) - \cos(\theta - \beta) + \cos\theta - \cos(\theta - \phi)$$

$$= 2 \sin \left\{\theta - \frac{1}{2}(\alpha + \beta)\right\} \sin \frac{1}{2}(\alpha - \beta) - 2 \sin\left(\theta - \frac{1}{2}\phi\right) \sin \frac{1}{2}\phi$$

$$= \pm 2 \sin \left\{\theta - \frac{1}{2}(\alpha + \beta)\right\} \cos \frac{1}{2}\phi - 2 \sin\left(\theta - \frac{1}{2}\phi\right) \sin \frac{1}{2}\phi$$

[using (5)]

Dividing by $\sin \frac{1}{2}\phi$ and transposing, we get

$$\frac{(l + l')}{r} \sec \frac{1}{2}\phi + 2 \cos\left(\theta - \frac{1}{2}\phi\right) \tan \frac{1}{2}\phi$$

$$= \pm 2 \cos \left\{\theta - \frac{1}{2}(\alpha + \beta)\right\}. \qquad ...(7)$$

Squaring (6) and (7) and then adding, we get

$$\left[\frac{l + l'}{r} \operatorname{cosec} \frac{\phi}{2} - 2 \cos\left(\theta - \frac{\phi}{2}\right) \cot \frac{\phi}{2}\right]^2$$

$$+ \left[\frac{l + l'}{r} \sec \frac{\phi}{2} + 2 \sin\left(\theta - \frac{\phi}{2}\right) \tan \frac{\phi}{2}\right]^2 = 4$$

This is the equation of the required locus. This represents conic, since if it is changed into cartesian coordinates we get a second degree equation in x and y.

2

Equation of the Conics Section

INTRODUTION

The equation of conic section is given by the general equation of the second degree in x and y which is

$$ax^2 + 2hxy + by^2 + 2gx + 2fy + c = 0. \quad ...(1)$$

Let the left hand side of (1) be denoted by F (x, y). Then the equation (1) becomes as

$$F(x, y) = 0. \quad ...(2)$$

If x, y are replaced by x_1, y_1 respectively in the left hand side of (2) then it will be denoted by

$$F(x_1, y_1)$$

Sometimes the equation (1) is also written as S = 0, and then the expression $F(x_1, y_1)$ is denoted by S_1.

CONIC THROUGH FIVE POINTS

The general equation of a conic, F (x, y) = 0, seems to have six constants namely a, b, c, f, g, h, but really it contains five independndent constants. On dividing the equation by any of these constants, say by c, the five independent constants are a/c, h/c, b/c, g/c and f/c.

Hence, five conditions are required to determine these five constants. For example, a conic can be made to pass through five given points.

Intersection of a Straight Line and a Conic

The equation of a straight line through a point (x_1, y_1) making an angle θ with the x-axis is

$$y - y_1 = \tan\theta\,(x - x_1)$$

$$\Rightarrow \quad \frac{x - x_1}{\cos\theta} = \frac{y - y_1}{\sin\theta}$$

$$\Rightarrow \quad \frac{x - x_1}{l} = \frac{y - y_1}{m} = r, \qquad ...(1)$$

where $l = \cos\theta$, $m = \sin\theta$ and r is the distance of any point (x, y) on the straight line (1) from the point (x_1, y_1).

The co-ordinates of any point P on (1) are $(lr + x_1, mr + y_1)$.

If the line (1) cuts the conic $F(x, y) = 0$ at the point P, then the point $P(lr + x_1, mr + y_1)$ will satisfy the equation $F(x, y) = 0$ and then we have

$$F(lr + x_1\ mr + y_1) = 0$$

$$\Rightarrow \quad a(lr + x_1)^2 + 2h(lr + x_1)(mr + y_1) + b(mr + y_1)^2 + 2g(lr + x_1) + 2f(mr + y_1) + c = 0$$

[$\because$ $F(x_1, y) = 0$ represents the equation (1) of δ 1]

$\Rightarrow$ $(al^2 + 2hlm + bm^2)r^2 + 2\{(ax_1 + hy_1 + g)l + (hx_1 + by_1 + f)m\}$ $r + F(x_1\ y_1) = 0,$ ...(2)

where $F(x_1, y_1) = ax_1^2 + 2hx_1 y_1 + by_1^2 + 2gx_1 + 2fy_1 + c.$

The equation (2), being a quadratic in r, gives two values of r.

TANGENT TO THE GENERAL CONIC

To find the equation of the tangent at any point (x_1, y_1) on the conic $F(x, y) = 0$.

Let the equation of the conic be given as

$$F(x, y) \equiv ax^2 = 2hxy + by^2 + 2gx + 2fy + c = 0. \qquad ...(1)$$

The equation of any straight line passing through the point (x_1, y_1) and making an angle θ with the x-axis is

$$y - y_1 = \tan\theta\,(x - x_1)$$

$$\Rightarrow \quad \frac{x - x_1}{\cos\theta} = \frac{y - y_1}{\sin\theta}$$

$$\Rightarrow \quad \frac{x - x_1}{l} = \frac{y - y_1}{m} = r.$$

where $l = \cos\theta$, $m = \sin\theta$ and r is the distance of any point P (x, y) on the straight line (2) from the point (x_1, y_1).

The coordinates of any point P (x, y) on the line (2) are given as

$(lr + x_1, mr + y_1)$. If the point $(lr + x_1, mr + y_1)$ lies on the conic (1), we have

$$a(lr + x_1)^2 + 2h(lr + x_1)(mr + y_1) + b(mr + y_1)^2 + 2g(lr + x_1) + 2f(mr + y_1) + c = 0$$

$$\Rightarrow (al^2 + 2hlm + bm^2) r^2 + 2r \{(ax_1 + hy_1 + g) l + (hx_1 + by_1 + f) m\} + F(x_1, y_1) = 0$$

$$\Rightarrow (al^2 + 2hlm + bm^2) r^2 + 2r \{(ax_1 + hy_1 + g) l + (hx_1 + by_1 + f) m\} = 0$$

[$\because$ the point (x_1, y_1) lies on the conic $F(x, y) = 0$ implies that $F(x_1, y_1) = 0$].

The quadratic (3) in r^2 gives the distances of the points of intersection of the straight line (2) and the conic (1) from the point (x_1, y_1). The straight line (2) will be a tangent to (1) at the point (x_1, y_1) if an only if it meets (1) at two coincident points at (x_1, y_1).

Clearly one root of the quadratic (3) in r^2 is zero. If the line (2) is to be a tangent, the other root of (3) should also be zero.

For this the coefficient of r in (3) should be zero.

i.e., $(ax_1 + hy_1 + g)\, l + (hx_1 + by_1 + f)\, m = 0.$ ...(4)

This is the condition that the line (2) will touch the conic $F(x, y) = 0$.

Eliminating l and m between (2) and (4), the equation of the tangent at the point (x_1, y_1) on the conic $F(x_1, y) = 0$ is given by

$$(ax_1 + hy_1 + g)(x - x_1) + (hx_1 + by_1 + f)(y - y_1) = 0$$

$$\Rightarrow x(ax_1 + hy_1 + g) + y(hx_1 + by_1 + f) = ax_1^2 + 2hx_1y_1 + by_1^2 + gx_1 + fy_1$$

$$\Rightarrow x(ax_1 + hy_1 + g) + y(hx_1 + by_1 + f) + gx_1 + fy_1 + c = ax_1^2 + 2hx_1y_1 + by_1^2 + 2gx_1 + 2fy_1 + c. \quad ...(5)$$

Since the point (x_1, y_1) lies on the conic (1), therefore

$$ax_1^2 + 2hx_1y_1 + by_1^2 + 2gx_1 + 2fy_1 + c = 0.$$

So the equation (5) becomes

$$x(ax_1 + hy_1 + g) + y(hx_1 + 2by_1 + f) + gx_1 + fy_1 + c) = 0. \quad ...(6)$$

$$\Rightarrow \quad x\left(\frac{1}{2}\frac{\partial F}{\partial x}\right)_{(x1,\, y1)} + y\left(\frac{1}{2}\frac{\partial F}{\partial y}\right)_{(x1,\, y1)} + c_1 = 0, \quad ...(7)$$

where $c_1 = gx_1 + fy_1 + c$

The equation (6) or (7) is the equation of the tangent to (1) at the point (x_1, y_1). It may also be written as

$$axx_1 + h(x_1y + xy_1) + hyy_1 + g(x + x_1) + f(y + y_1) + c = 0 \quad ...(8)$$

Working Rule : To write the equation of then tangent at any point (x_1, y_1) on a conic, we write in the equation of the conic :

(i) xx_1 for x^2,

(ii) yy_1 for y^2,

(iii) $x_1y + xy_1$ for $2xy$,

(iv) $x + x_1$ for $2x$,

(v) $y + y_1$ for $2y$ and

(vi) retain the constant term

CONDITION OF TANGENCY

To find the condition that the line $lx + my + n = 0$ may touch the conic $F(x, y) = 0$

Let the given line be

$$lx + my + n = 0 \quad ...(1)$$

This line touches the conic $F(x, y) = 0$ at the point (x_1, y_1). The equation of the tangent at the point (x_1, y_1) on the conic $F(x, y) = 0$ is

$$x(ax_1 + hy_1 + g) + y(hx_1 + by_1 + f) + (gx_1 + fy_1 + c) = 0 \quad ...(2)$$

The equations (1) and (2) respresent the same straight line.

Therefore comparing the coefficient, we have

$$\frac{ax_1 + hy_1 + g}{l} = \frac{hx_1 + by_1 + f}{m} = \frac{gx_1 + fy_1 + c}{n} = \lambda, \text{ (say)}.$$

Therefore

$$ax_1 + hy_1 + g - l\lambda = 0, \quad ...(3)$$

$$hx_1 + by_1 + f - m\lambda = 0, \quad ...(4)$$

and $gx_1 + fy_1 + c - n\lambda = 0.$ ...(5)

Also (x_1, y_1) lies on the line (1). Therefore

$$lx_1 + my_1 + n = 0$$

$$\Rightarrow \quad lx_1 + my_1 + n = 0.\ \lambda = 0. \quad ...(6)$$

Eliminating x_1, y_2 and λ between the four relations (3), (4), (5) and (6), the required condition is

$$\begin{vmatrix} a & h & g & l \\ h & b & f & m \\ g & f & c & n \\ l & m & n & 0 \end{vmatrix} = 0.$$

Expanding the above determinant, we have

$$Al^2 + Bm^2 + Cn^2 + 2\ Fmn + 2Gnl + 2Hlm = 0$$

where A, B C, F, G, H are the cofactors of A, b, c, f, g, h in the determinant

$$\begin{vmatrix} a & h & g \\ h & b & f \\ g & f & c \end{vmatrix}. \quad \textbf{[Remember]}$$

DIRECTOR CIRCLE OF A CONIC

Definition: *The locus of the point of intersection of perpendicular tangents, is called a director circle.*

To find the equation of the director circle of a conic

Let the equation of the conic be $F\ (x, y) = 0$.

Now let (x_1, y_1) be the point of intersection of any two perpendicular tangents to the conic $F\ (x, y) = 0$.

The equation of the pair of tangents from (x_1, y_1) to the conic $F\ (x, y) = 0$ is given by

$$SS_1 = T^2$$

$$\Rightarrow \quad (ax^2 + 2hxy + by^2 + 2gx + 2fy + c) \times (ax_1^2 + 2hx_1y_1 + by_1^2 + 2gx_1 + 2fy_1 + c)$$

$$- \{x\ (ax_1 + hy_1 + g) + y\ (hx_1 + by_1 + f) + gx_1 + fy_1 + c\}^2 = 0. \quad ...(1)$$

The two tangents given by (1) will be perpendicular to each other if in the equation (1), we have the coeff. of x^2 + the coeff, of $y^2 = 0$

$$\Rightarrow \quad (a + b)\ (ax_1^2 + 2hx_1y_1 + by_1^2 + 2gx_1 + 2fy_1 + c) - (ax_1 + hy_1 + g)^2 - (hx_1 + by_1 + f)^2 = 0 \quad ...(2)$$

$\therefore$ the locus of (x_1, y_1) *i.e.*, the equation of the director circle is given by

$$(ab - h^2)(x^2 + y^2) + 2x(bg - hf) + 2y(af - gh) + c(a + b) - g^2 - f^2 = 0$$

$$\Rightarrow \quad C(x^2 + y^2) - 2Gx - 2Fy + (A + B) = 0 \qquad ...(3)$$

where A, B, C, F, G, H have same values as

Particular Case: *If $ab - h^2 = 0$, the conic $F(x, y) = 0$ respresents a parabola. The equation (3) of director circle reduces to*

$$2Gx + 2Fy - (A + B) = 0. \qquad [\because ab - h^2 = 0 \text{ means } C = 0]$$

This is the equation of the directrix of the parabola.

Foci

To show that a central conic has four and only four foci, two of which are real and two imaginary.

Let the equation of a central conic be

$$ax^2 + by^2 = 1. \qquad ...(1)$$

Let the coordinates of its focus be (x_1, y_1) and the equation of the corresponding directrix be

$$x \cos\alpha + y \sin\alpha = p = 0. \qquad ...(2)$$

The distance of a point (x, y) on it from the focus = e × the perpendicular distance of (x, y) from the directrix

$$\Rightarrow \quad \sqrt{\{(x - x_1)^2 + (y - y_1)^2\}}$$

$$= e(x \cos\alpha + y \sin\alpha - p)/\sqrt{\{(\cos^2\alpha + \sin^2\alpha)}.$$

Squaring, we get

$$(x - x_1)^2 + (y - y_1)^2 = e^2\{x^2\cos^2\alpha + y^2\sin^2\alpha + p^2 - 2px\cos\alpha$$

$$-2py\sin\alpha + 2xy\cos\alpha\sin\alpha\}$$

$$\Rightarrow \quad x^2(1 - e^2\cos^2\alpha) + y^2(1 - e^2\sin^2\alpha) - 2e^2\cos\alpha\sin\alpha\, xy$$

$$-2x(x_1 - pe^2\cos\alpha) - 2y(y_1 - pe^2\sin\alpha)$$

$$+ (x_1^2 + y_1^2 + p^2e^2) = 0 \qquad ...(3)$$

The equations (1) and (3) should represent the same conic, therefore comparing their coefficients, we have

$$\frac{1 - e^2\cos^2\alpha}{a} = \frac{1 - e^2\sin^2\alpha}{b} = \frac{2e^2\cos\alpha.\sin\alpha}{0} = \frac{x_1 - pe^2\cos\alpha}{0}$$

$$= \frac{y_1 - pe^2\sin\alpha}{0} = \frac{x_1^2 + y_1^2 - p^2e^2}{-1} \qquad ...(4)$$

From the third ratio, we have

$\cos\alpha \sin\alpha = 0 \Rightarrow \cos\alpha = 0$ or $\sin\alpha = 0$

$\Rightarrow \alpha = \frac{1}{2}\pi$ or $\alpha = 0$

Let us first take $\alpha = 0$.

From the 4th ratio, we have

$$x_1 = pe^2 \cos\alpha \Rightarrow x_1 = pe^2 \quad ...(5)$$

From the 5th ratio, we have

$$y_1 = pe^2 \sin\alpha \Rightarrow y_1 = 0 \quad ...(6)$$

From the 1st and 2nd ratios, we have

$$\frac{1-e^2}{a} = \frac{1}{b} \Rightarrow e = \left(1 - \frac{a}{b}\right)^{1/2} \quad ...(7)$$

From the 1st and last ratios, we have

$$\frac{1-e^2}{a} = \frac{x_1^2 + 0 - e^2p^2}{-1} \quad \text{(using (6))}$$

$$\Rightarrow e^2 - 1 = a\,\{p^2e^2 - e^2p^2\} \quad \text{(using (5) for } x_1)$$

$$\Rightarrow (e^2 - 1) = ap^2e^2\,(e^2 - 1\}$$

$$\Rightarrow 1 = ap.\,pe^2 \Rightarrow apx_1 = 1. \quad ...(8)$$

From the 2nd and last ratios,

$$\frac{1}{b} = \frac{x_1^2 + p^2e^2}{-1} \quad \text{(using (6))}$$

$$\Rightarrow x_1^2 = e^2p^2 - \frac{1}{b} = \frac{x_1^2}{e^2}\frac{1}{b} \quad \text{(using (5) for p)}$$

$$\Rightarrow x_1^2\left(\frac{1}{e^2} - 1\right) = \frac{1}{b}, \text{ or } x_1^2\left(\frac{b}{b-a} - 1\right) = \frac{1}{b} \quad \text{(using (7))}$$

$$\Rightarrow x_1^2\,\frac{1}{a} - \frac{1}{b} \quad ...(9)$$

From (9), we see that there are two foci on the x-axis whose distances from the centre (0, 0) are $\pm\sqrt{\left(\frac{1}{a} - \frac{1}{b}\right)}$.

Now a focus on x-axis is $(x_1, 0)$. Its polar w.r.t. (1) is

$$axx_1 + 0 = 1 \text{ or } x.\frac{1}{p} = 1. \quad [\because \text{ from (8), } apx_1 = 1]$$

$$\Rightarrow \quad x = p.$$

Also when $\alpha = 0$, the equation (2) of directrix is $x = p$.

This shows that the directrix is the polar of the corresponding focus.

Now 1st us take $\alpha = \pi/2$. Discussing as above, we have

$$x_1 = 0,\ y_1 = e^2 p,\ e = \sqrt{\left(1 - \frac{a}{b}\right)},\ bpy_1 = 1,\ \ y_1^2 = \frac{1}{b} - \frac{1}{a}.$$

From the last relation it follows that there are two other foci on y-axis whose distances from the centre (0, 0) are $\pm\sqrt{\left(\frac{1}{b} - \frac{1}{a}\right)}$.

Two pairs of foci are $(\pm\, x_1, 0)$ and $(0, \pm\, y_1)$ where $x_1^2 = 1/a - 1/b$ and $y_1^2 = 1/b - 1/a$. Clearly one and only one pair of foci is real and the other imaginary whatever real values *a* and *b* may take. We have an important relation:

If e_1 and e_2 be the two eccentricities, then

$$\frac{1}{e_1^2} + \frac{1}{e_2^2} = \frac{b}{b-a} + \frac{a}{a-b} = 1.$$

TANGENTS FROM A FOCUS TO A CONIC

To show that the tangents from a focus to a conic satisfy the conditions for a circle.

Let (0, 0) be a focus of the conic and let the equation of its directrix be

$$x \cos \alpha + y \sin \alpha - p = 0.$$

By definition, the equation of the conic is given by

$$(x - 0)^2 + (y - 0)^2 = e^2 \left\{ \frac{x \cos \alpha + y \sin \alpha - p}{\sqrt{(\cos^2 \alpha + \sin^2 \alpha)}} \right\}^2$$

$$\Rightarrow \quad x^2 + y^2 - e^2 (x \cos \alpha + y \sin \alpha - p)^2 = 0. \qquad ...(1)$$

$$\Rightarrow \quad (x + iy)(x - iy) - e^2 (x \cos \alpha + y \sin \alpha - p)^2 = 0. \qquad ...(2)$$

The equation (2) shows that both the imaginary lines $x \pm iy = 0$ meet the conic (2) in coincident points which are given by $(x \cos \alpha + y \sin \alpha - p)^2 = 0$. Hence the tangents from a focus to a conic are the imaginary lines $x \pm iy = 0$ whose combined equation is

$$(x + iy)(x - iy) = 0 \quad \textit{i.e.,} \quad x^2 + y^2 = 0,$$

which is a point circle at the focus (0, 0). We have proved in § 13 above that the directrix is the polar of the focus and hence the chord of contact of these tangents (imaginary tangents) is the directrix

$$x \cos \alpha + y \sin \alpha - p = 0.$$

Now if the origin is shifted and the axes are rotated in any manner, the equation of tangents from a focus namely the equation $x^2 + y^2 = 0$ can take the form

$$x^2 + y^2 + 2gx + 2fy + c = 0.$$

The last equation is the equation of a circle and hence the proposition is proved.

TO FIND THE FOCI OF A CONIC

(A) *Let the equation of a conic be*

$$F(x, y) = 0 \qquad ...(1)$$

Let (x_1, y_1) be the focus. Then the equation of an imaginary line thought he focus is given by

$$(x - x_1) + i(y - y_1) = 0$$

$$\Rightarrow \quad x + iy - (x_1 + iy_1) = 0. \qquad ...(2)$$

Comparing (1) with $lx + my + n = 0$, we have

$$l = 1, m = l, n = -(x_1 + iy_1).$$

The line (2) will be tangent to the conic (1), if the condition namely

$$Al^2 + Bm^2 + Cn^2 + 2Fmn + 2Gnl + 2Hlm = 0 \text{ is satisfied.}$$

$$\therefore \quad A(1)^2 + B(i)^2 + C(x_1 + iy_1)^2 - 2Fi(x_1 + iy_1)$$
$$-2G(x_1 + iy_1).\, l + 2H.l.i. = 0$$

$$\Rightarrow \quad \{A - B + C(x_1^2 - y_1^2) + 2Fy_1 - 2Gx_1\}$$
$$+ 2i\{Cx_1y_1 - Fx_1 - Gy_1 + H\} = 0.$$

Separating real and imaginary parts, we have

$$A - B + C(x_1^2 - y_1^2) + 2Fy_1 - 2Gx_1 = 0. \qquad ...(2)$$

and $Cx_1y_1 - Fx_1 - Gy_1 + H = 0.$ ...(3)

The equations (2) and (3) give the coordinates (x_1, y_1) of the focus.

(B) *Another form*

Let (x_1, y_1) be the coordinates of a focus of the conic

$$F(x, y) = 0. \qquad ...(1)$$

The equation to the pair of tangents from (x_1, y_1) to the conic is

$$\Rightarrow \quad \{ax^2 + 2hxy + by^2 + 2gx + 2fy + c\} F(x_1, y_1)$$
$$= \{x(ax_1 + hy_1 + g) + y(hx_1 + by_1 + f) + gx_1 + fy_1 + c\}^2.$$

This equation will satisfy the conditions of a circle.

$\therefore$ the coeff. of x^2 = the coeff. of y^2, and the coeff, of $xy = 0$.

Thus we have

$$aF(x_1, y_1) - (ax_1 + hy_1 + g)^2 = F(x_1 + y_1) - (hx_1 + by_1 + f\}^2$$

and $hF(x_1, y_1) = (ax_1 + hy_1 + g)(hx_1 + by_1 + f)$.

Therefore, the foci are given by

$$\frac{(ax_1 + hy_1 + g)^2 - (hx_1 + by_1 + f)^2}{a - b} = \frac{(ax_1 + hy_1 + g)(hx_1 + by_1 + f)}{h}$$
$$= F(x_1, y_1). \qquad ...(4)$$

CONJUGATE LINES

Definition: *Two lines are called conjugate lines if the pole of one lies on he other.*

To find the condition that the lines $l_1x + m_1y + n_2 = 0$ and $l_2x + m_2y + n_2 = 0$ are conjugate lines.

Let the equation of the conic be given as

$$F(x, y) = 0. \qquad ...(1)$$

Let the pole of the line

$$l_1x + m_1y + n_1 = 0 \qquad ...(2)$$

w.r.t. the conic (1) be (x_1, y_1).

The polor of (x_1, y_1) w.r.t. the conic (1) is given by $T = 0$

i.e., $x(ax_1 + hy_1 + g) + y(hx_1 + by_1 + f) + gx_1 + fy_1 + c = 0.$ ...(3)

The equations (2) and (3) represent the same straight line. Therefore comparing their coefficients, we get

$$\frac{ax_1 + hy_1 + g}{l_1} = \frac{hx_1 + by_1 + f}{m_1} = \frac{gx_1 + fy_1 + c}{n_1} = \lambda, \text{ say.}$$

Which gives

$$ax_1 + hy_1 + g - l_1\lambda = 0,$$
$$hx_1 + by_1 + f - m_1\lambda = 0,$$

and $gx_1 + fy_1 + c - n_1\lambda = 0,$

Since $l_2x + m_2y + n_2 = 0$ passes through (x_1, y_1), therefore

we have $\quad l_2x_1 + m_2y_1 + n_2 = 0,\ l = 0$

Eliminating x_1, y_1 and λ between the above four relations, we have

$$\begin{vmatrix} a & h & g & l_1 \\ h & b & f & m_1 \\ g & f & c & n_1 \\ l_2 & m_2 & n_2 & 0 \end{vmatrix} = 0, \text{ which is the required condition}$$

Expanding this determinant

$$Al_2l_2 + Bm_1m_2 + Cn_1n_2 + F(m_1n_2 + m_1n_2) + G(n_1l_1 + n_2l_1) + H(l_1m_2 + l_2m_2) = 0$$

where A, B, C, F, G, H are the cofactors of a, b, c, f, g, h in the determinant

$$\begin{vmatrix} a & h & g \\ h & b & f \\ g & f & c \end{vmatrix}.$$

CHORD WITH A GIVEN MIDDLE POINT

Let the equation of the conic be given as

$$F(x, y) \equiv ax^2 + 2hxy + by^2 + 2gx + 2fy + c = 0. \qquad ...(1)$$

Let the middle point of the chord be (x_1, y_1) and suppose that the equation of the chord is given as where r is the distance of any point (x, y) on (2) from the point (x_1, y_1).

$$\frac{x - x_1}{l} = \frac{y = y_1}{m} = r \text{ (say)} \qquad ...(2)$$

Where r is the distance of any point (x, y) on (2) from the point (x_1, y_1)

Any point on (2) at a distance r from (x_1, y_1) is

$$(x_1 + lr,\ y_1 + mr).$$

The two values of r for the points of intersection of (2) with the conic (1) are given by

$$r^3 (al^2 + 2hlm + bm^2) + 2r\{(ax_1 + hy_1 + g)\,1 + (hx_1 + by_1 + f)\,m\} + F(x_1, y_1) = 0$$

Since (x_1, y_1) is the middle point, therefore the two values of r, say r_1 and r_2, should be equal in magnitude but opposite in sign.

$$\therefore \qquad r_1 + r_2 = 0$$

$$\Rightarrow \quad -\frac{\text{coeff. of r}}{\text{coeff. of } r^2} = 0 \text{ coeff. or r} = 0$$

$$\Rightarrow \quad (ax_1 \mid hx_1 + g)\, l + (xh_1 + by_1 + f)\, m = 0. \quad \text{...(3)}$$

Eliminating l, m between (2) and (3), the required equation of the chord with (x_1, y_1) as its middle point is given by

$$(ax_1 + hy_1 + g)(x - x_1) + (hx_1 + by_1 + f)(y - y_1) = 0$$

$$\Rightarrow \quad x(ax_1 + hy_1 + g) + y(hx_1 + by_1 + f) = ax_1^2 + 2hx_1y_1 + by_1^2 + gx_1 + fy_1$$

$$\Rightarrow \quad x(ax_1 + hy_1 + g) + y(hx_1 + by_1 + f) + gx_1 + fy_1 + c = ax_1^2 + 2hx_1y_1 + by_1^2 + 2gx_1 + 2fy_1 + c$$

$$\Rightarrow \quad axx_1 + h(xy_1 + x_1y) + byy_1 + g(x + x_1) + f(y + y_1) + c = ax_1^2 + 2hx_1y_1 + by_3^2 + 2gx_1 + 2fy_1 + c$$

$\Rightarrow \quad T = S_1$, where

$T \equiv axx_1 + h(xy_1 + x_1y) + byy_1 + g(x + x_1) + f(y + y_1) + c$ and S_1 is what $F(x, y)$ becomes when x and y are replaced by x_1 and y_1.

Diameter

Definition: *The locus of the middle points of a system of parallel chords is called a diameter.*

The equation of the diameter. To find the equation of a diameter of the conic $F(x, y) = 0$.

Let the equation of the conic is $F(x, y) = 0$. ...(1)

Let the equation of a system of parallel chords of the conic (1) be given by

$$y = mx + \lambda. \quad \text{...(2)}$$

where m is a constant and λ a parameter.

Let (x_1, y_1) be the co-ordinates of the middle point of a chord given by the system (2). The equation of the this chord is then given by

$$T = S_1$$

$$\Rightarrow \quad x(ax_1 + hy_1 + g) + y(hx_1 + by_1 + f) + gx_1 + fy_1 + c = ax_1^2 + 2hx_1y_1 + by_1^2 + 2gx_1 + 2fy_1 + c \quad \text{...(3)}$$

The straight line given by (3) belongs to the system of parallel lines (2).

$\therefore$ the gradient of (3) = the gradient of (2)

$\Rightarrow$ $-(ax_1 + hy_1 + g)/(hx_1 + by_1 + f) = m$

$\Rightarrow$ $(ax_1 + hy_1 + g) + m\,(hx_1 + by_1 + f) = 0$

$\Rightarrow$ $(a + mh)\,x_1 + (h + bm)\,y_1 + g + mf = 0$

$\therefore$ the locus of the middle point (x_1, y_1) is given by

$$(a + hm)\,x + (h + bm)\,y + g + mf = 0. \qquad ...(4)$$

CONJUGATE DIAMETERS

Definition : *If two diameters are such that each diameter bisects the chords parallel to the other, they are called conjugate diameters.*

We know that the equation of the diameter bisecting the chords parallel to $y = mx$ is given by

$$(a + mh)\,x + (h + bm)\,y + g + mf = 0.$$

If this diameter is parallel to $y = m'x$, then

$$m' = -(a + mh)/(h + mb)$$

$\Rightarrow$ $a + h\,(m + m') + bmm' = 0$

Which is the condition that $y = mx$ and $y = m'x$ be parallel to conjugate diameters of the conic $F\,(x, y) = 0$.

FIND CONDITION THAT THE TWO STRAIGHT LINES

$$Ax^2 + 2Hxy + By^2 = 0$$

may be conjugate diameters of the conic

$$ax^2 + 2hxy + by^2 = 1$$

The equation of the conic is given by

$$ax^2 + 2hxy + by^2 = 1 \qquad ...(1)$$

The equation of the pair of straight lines is given as

$$Ax^2 + 2Hxy + By^2 = 0 \qquad ...(2)$$

Now let the equations of the two straight lines given by (2) be

$$y = mx$$

and $y = m'x$, so that

$$B\,(y - mx)\,(y - m'x) \equiv Ax^2 + 2Hxy + By^2.$$

$\therefore$ $m + m' = \;2H/B$ and $mm' = A/B$. ...(3)

Now if the lines $y = mx$ and $y = m'\,x$ be conjugate diameters of the conic (1) then using the condition

$$a + h(m + m') + bmm' = 0,$$

we have $a + h(-2H/B) + b(A/B) = 0$

i.e., $aB - 2hH + bA = 0$. This is required condition.

To show that any two concentric conics have in general on and only one pair of common conjugate diameters. Hence to show that the equation of any two concentric conics can be put in the form

$$ax^2 + by^2 = 1$$

and $a'x^2 + b'y^2 = 1$.

Let the equations of two concentric conics with centre (0, 0)

be given as $ax^2 + 2hxy + by^2 = 1$

and $a'x^2 + 2h'xy + b'y^2 = 1$.

Now let the equations of a pair of common conjugate diameters be given by

$$Ax^2 + 2Hxy + By^2 = 0 \quad ...(1)$$

Now the diameters given by (1) are conjugate diameters w.r.t. both the conics if

$$aB - 2hH + bA = 0$$

and $a'B - 2h'H + b'A = 0$

Solving these equations, we have

$$\frac{B}{2(bh' - b'h)} = \frac{H}{a'b - ab'} = \frac{A}{2(a'h - ah')} \quad ...(2)$$

Eliminating B, H, A between (1) and (2), the equation of the common conjugate diameters of the conics is

$$(a'h = ah')x^2 + (a'b = ab')xy + (bh' - b'h)y^2 = 0$$

Now taking common conjugate diameters as oblique axes, the equations of the two conics can be reduced to the form

$$ax^2 + by^2 = 1, \text{ and } a'x^2 + b'y^2 = 1.$$

PAIR OF TANGENTS

To find the equation of the pair of tangents from a point (x_1, y_1) to the conic $F(x, y) = 0$.

Let the equation of the conic is $F(x, y) = 0$

The equation of any line through the point (x_1, y_1) is

$$\frac{x - x_1}{l} = \frac{y - y_1}{m} = r$$

Any point ($lr + x_1$, $mr + y_1$) on (1) will lie on the conic $F(x, y) = 0$ if

$$r^2 (al^2 + 2hlm + bm^2) + 2r \{(ax_1 + hy_1 + g)\, l + (hx_1 + by_1 + f)\, m\} + F(x_1, y) = 0. \quad ...(2)$$

If the line (1) is a tangent to the conic $F(x, y) = 0$, then the two values of r given by (2) are coincident so that we have the discriminant of (2) = 0

i.e., $$4 \{(ax_1 + hy_1 + g)\, l + (hx_1 + by_1 + f)\, m\}^2 - 4 (al^2 + 2hlm + bm^2)\, F(x_1, y_1) = 0 = 0. \quad ....(3)$$

Putting the values of l, m from (1) in (3), the required equation of the pair of tangents is given by

$$\{(ax_1 + hy_1 + g)(x - x_1) + (hx_1 + by_1 + f)(y - y_1)\}^2 = F(x_1 + y_1)\{a(x - x_1)^2 + 2h(x - x_1)(y - y_1) + b(y - y_1)^2\}.$$

This equation can be put in the form

$$SS_l = T^2$$

where S, S_1, T have their usual meanings.

AXES OF THE CONIC

From above we have

$$\frac{(ax + hy + g)^2 - (hx + by + f)^2}{a - b} = \frac{(ax + hy + g)\ (hx + by + f)}{h} \quad ...(1)$$

represents some conic passing through the four foci of the conic F (x, y) =0 as well as through the centre. Note that the centre is given by the following equation.

$$\frac{\partial F}{\partial x} = 0$$

i.e., $ax + hy + g = 0$

and $\dfrac{\partial F}{\partial y} = 0$ *i.e.,*

$$hx + by + f = 0.$$

But we know that the only conic passing through the four foci and the centre is the pair of axes of the conic. Hence the equations of the axes are given by the equation (1).

To find the general equation of the conic passing through the vertices of a quadrilateral

Let the equations of the sides of the quadrilateral taken in order be $u_1 = 0$, $u_2 = 0$, $u_3 = 0$, $u_4 = 0$. Hence $u_1u_3 = 0$ is a conic passing through the four vertices and $u_2u_4 = 0$ is another conic passing through them.

Thus $u_1u_3 = \lambda u_2u_4$ is the general equation of a conic passing thought he four vertices.

To find the equation of the circle of curvature at a point on the conic $S = 0$.

Let the equation of the conic be

$$S \equiv ax^2 + 2hxy + by^2 + 2gx + 2fy + c = 0.$$

The equation of the tangent to the conic $S = 0$ at any point (x_1, y_1) on it is

$$T \equiv axx_1 + h\,(x_1y + xy_1) + byy_1 + g\,(x + x_1) + f\,(y + y_1)\,c = 0.$$

Again let

$$u \equiv lx + my + n = 0$$

be the equation of any straight line passing thought he point (x_1, y_1).

Then the equation $S + \lambda.u.T = 0$ represents a conic which has a contact of the second order with $S = 0$ at (x_1, y_1) on it. If now λ is so chosen that $S + \lambda.u.T = 0$ is a circle, it will be the equation of the circle of curvature at (x_1, y_1).

EQUATION OF A CONIC REFERRED TO TANGENT AND NORMAL AS COORDINATE AXES

Since the tangent and normal at a point on the conic are taken as the coordinate axes, therefore the point of constant is the origin. Thus the origin must lie on the conic and, therefore, the equation of the conic will not contain the constant term. So its equation can be written as

$$ax^2 + 2hxy + by^2 + 2gx + 2fy = 0. \qquad ...(1)$$

Again let us choose x-axis along the tangent, so that $y = 0$ meets (1) in two coincident points at origin given by

$$ax^2 + 2gx = 0.$$

The roots of this equation will be coincident if its discriminant is zero *i.e.*, $4g^2 = 0$ *i.e.*, $g = 0$.

Putting $g = 0$ in the equation (1), the required equation of the conic is given by

$$ax^2 + 2hxy + by^2 + 2fy = 0$$

CONFOCAL CONICS

Definition: *All those conics which have the same two points as their foci are called confocal conics.*

The equation

$$\frac{x^2}{a^2 + \lambda} + \frac{y^2}{b^2 + \lambda} = 1$$

where λ is a parameter, represents a family of confocals. Assuming that a < b, the foci are the points.

$$[\pm \sqrt{\{(a^2 + l) - (b^2 + \lambda)\}}, 0] \quad i.e., \{\pm \sqrt{\{(a^2 - b^2)}, 0\},$$

which remain the same, whatever λ may be. The same can be shown to be true if a < b.

TO FIND THE EQUATION OF CONFOCALS TO AN ELLIPSE

Taking the axes of the ellipse as the coordinate axes and the centre as origin, the equation of the ellipse (conic) is given by

$$x^2/a^2 + y^2/b^2 = 1. \qquad ...(1)$$

We know that all conics which have the same foci have the same centre and also their axes are in the same directions.

Hence the equation of a conic having the same centre and the same axes as the original conic may be taken as

$$x^2/\alpha^2 + y^2/\beta^2 = 1. \qquad ...(2)$$

The real foci of (1) are $(\pm \sqrt{(a^2 - b^2)}, 0)$ and those of (2) are $(\pm \sqrt{(\alpha^2 - \beta^2)}, 0)$.

Since (1) and (2) are confocals, they have the same foci.

$$\therefore \quad \sqrt{(a^2 - b^2)} = \sqrt{(\alpha^2 - \beta^2)}.$$

or

$$a^2 - b^2 = \alpha^2 - \beta^2$$

$$\therefore \quad \alpha^2 = a^2 + \lambda \text{ and } \beta^2 = b^2 + \lambda.$$

Hence the equation of the system of conics confocal to (1) is

$$\frac{x^2}{a^2 + \lambda} + \frac{y^2}{b^2 + \lambda} = 1.$$

PROPOSITIONS ON CONFOCALS CONICS

(A) *Through any given point in the plane of a conic, two confocals can be drawn, one of which is a hyperbola and other an ellipse.*

Let the equation of the given conic be

$$x^2/a^2 + y^2/b^2 = 1. \qquad ...(1)$$

Let (h, k) be the co-ordinates of the given point. The equation of a conic confocal to (1) is

$$\frac{x^2}{a^2+\lambda} + \frac{y^2}{b^2+\lambda} = 1 \qquad ...(2)$$

If the conic (2) passes through the point (h, k), we have

$$\frac{h^2}{a^2+\lambda} + \frac{k^2}{b^2+\lambda} = 1 \qquad ...(3)$$

Since (3) is a quadratic equation in λ, therefore it gives two values of λ corresponding to which we have two confocals given by (2).

Let us put $b^2 + \lambda = \mu$; then $a^2 + \lambda = a^2 + \mu - b^2 = a^2e^2 + \mu$.

Hence (3) becomes

$$\frac{h^2}{a^2e^2+\mu} + \frac{k^2}{\mu} = 1$$

$\Rightarrow$
$$\mu^2 + (a^2e^2 - h^2 - k^2)\,\mu - a^2e^2k^2 = 0 \qquad ...(4)$$

The constant term namely $(-a^2e^2k^2)$ in (4) in negative. This shows that one value of μ *i.e.,* of $b^2 + \lambda$ is positive and the other negative. Similarly we can prove that both values of $a^2 + \lambda$ are positive. Hence of the two confocals one is an ellipse and the other is a hyperbola.

(B) *Confocals cut at right angles.*

Let the equations of the two confocals be

$$\frac{x^2}{a^2+\lambda_2} + \frac{y^2}{b^2+\lambda_2} = 1 \qquad ...(1)$$

and
$$\frac{x^2}{a^2+\lambda_2} + \frac{y^2}{b^2+\lambda_2} = 1 \qquad ...(2)$$

If (1) and (2) cut at the point (x_1, y_1), we have

$$\frac{x_1^2}{a^2+\lambda_1} + \frac{y_1^2}{b^2+\lambda_1} = 1 \qquad ...(3)$$

and
$$\frac{x_1^2}{a^2+\lambda_1} + \frac{y_1^2}{b^2+\lambda_1} = 1 \qquad ...(4)$$

Subtracting (4) from (3), we have

$$\frac{x_1^2\,(\lambda_2-\lambda_1)}{(a^2+\lambda_1)(a^2+\lambda_2)} + \frac{y_1^2\,(\lambda_2-\lambda_1)}{(b^2+\lambda_1)(b^2+\lambda_2)} = 0$$

$$\Rightarrow \qquad \frac{x_1^2}{(a^2 + \lambda_1)(a^2 + \lambda_2)} + \frac{y_1^2}{(b^2 + \lambda_1)(b^2 + \lambda_2)} = 0 \quad ...(5)$$

$$[\because \lambda_1 \neq \lambda_2]$$

The equations of the tangents to the two conics (1) and (2) at the point (x_1, y_1) are

$$\frac{xx_1}{a^2 + \lambda_1} + \frac{yy_1}{b^2 + \lambda_1} = 1 \quad ...(6)$$

and
$$\frac{xx_1}{a^2 + \lambda_2} + \frac{yy_1}{b^2 + \lambda_2} = 1 \quad ...(7)$$

The equations (6) and (7) will be at right angles to each other if

[the slope of (6)] × [the slope of (7)] = – 1

i.e., if
$$\left(-\frac{x_1}{a^2 + \lambda_1} \cdot \frac{b^2 + \lambda_1}{y_1}\right) \times \left(-\frac{x_1}{a^2 + \lambda_1} \cdot \frac{b^2 + \lambda_2}{y_1}\right) = -1$$

i.e., if
$$\frac{x_1^2}{(a^2 + \lambda_1)(a^2 + \lambda_2)} + \frac{y_1^2}{(b^2 + \lambda_1)(b^2 + \lambda_2)} = 0. \quad ...(8)$$

The condition (8) in true by virtue of (5). Hence the two confocals cut at right angles.

(C) *One and only one member of confocals touches a given line.*

Let the equation of a system of confocals be

$$\frac{x^2}{a^2 + \lambda} + \frac{y^2}{b^2 + \lambda} = 1, \quad ...(1)$$

and the equation of a given line be

$$lx + my + n = 0 \quad ...(2)$$

The line (2) will touch (1) if

$$(a^2 + \lambda) l^2 + (b^2 + \lambda) m^2 = n^2$$

which is a linear equation in λ and so gives one and only one value of λ. Hence the proposition.

SOLVED EXAMPLES

Example 1:

If two conics have their axes parallel, prove that a circle will pass through their points of intersection.

Solution:

Let the co-ordinate axes be chosen parallel to the axes of the given conics, so that the equations of the two conics may be taken as

$$ax^2 + by^2 + 2gx + 2fy + c = 0 \qquad ...(1)$$

and $$a'x^2 + b'y^2 + 2g'x + 2f'y + c' = 0 \qquad ...(2)$$

The general equation of a conic passing through the points of intersection of the conics (1) and (2) is given by

$$(ax2 + by^2 + 2gx + 2fy + c) + \lambda\,(a'x^2 + b'y^2 + 2g'x + 2f'y + c') = 0 \qquad ...(3)$$

If the equation (3) represents a circle, we have

the coefficient of x^2 = the coefficient of y^2

$$\Rightarrow \qquad (a + \lambda a^2) = b + \lambda b', \quad \text{or} \quad \lambda = (b - a)/(a' - b').$$

Hence the equation (3) represents the equation of a circle when $\lambda = (b - a)/(a' - b')$. This proves the required statement.

Example 2:

If a circle has double contact with a conic, the chord of contact is parallel to one or other of the axes.

Solution:

Referred to the axes of the conic as the coordinate axes let the equation of the conic be

$$ax^2 + by^2 = 1. \qquad ...(1)$$

Let the equation of the chord joining the points of contact of the conic and the circle be given by

$$lx + my + n = 0. \qquad ...(2)$$

The equation of a conic having double contact with the conic (1) at the points of intersection of (1) and (2) is given by

$$(ax^2 + by^2 - 1) + \lambda\,(lx + my + n)^2 = 0 \qquad ...(3)$$

The equation (3) will represent a circle if the coefficient of xy = 0 *i.e.*, $2\lambda lm = 0$ *i.e.*, $l = 0$ or $m = 0$, since $\lambda \equiv 0$

and the coefficient of x^2 = the coefficient of y^2

$$\Rightarrow \qquad a + \lambda^2 = b + \lambda m^2. \qquad ....(4)$$

If $l = 0$ the equation (2) of the chord of contact is given by $my + n = 0$, which is a straight line parallel to x-axis.

And if m = 0 the equation (2) of the chord of contact becomes $lx + n = 0$ and hence it is parallel to y-axis.

The value of λ corresponding to $l = 0$ or $m = 0$ is determined from the relation (4).

Hence the equation (3) will represent a circle if the chord of contact (2) of the conic (1) and the circle is parallel to one or the other axis of the conic (1).

Example 3:

If two circles have double contact with a conic, and the chords of contact are perpendicular, their point of intersection is limiting point of the co-axal system determined by the two circles.

Solution:

Proceed as above in Ex. 3.

If $l = 0$, then from relation (4), we get $\lambda = (a = b)/m^2$.

Putting this value of λ in the equation (3), the equation of one circle is given by

$$(ax^2 + by^2 - 1) + \{(a - b)/m^2\} (my + n)^2 = 0$$

$$\Rightarrow \quad ax^2 + by^2 - 1 + (a - b) (y + n/m)^2 = 0$$

$$\Rightarrow \quad ax^2 + by^2 - 1 + (a - b) (y + \alpha)^2 = 0. \qquad ...(5)$$

where $\alpha = -n/m$.

Again when m = 0 then from the relation (4), $\lambda = (b = a)/\lambda^2$.

Putting this value of λ in the equation (3), the equation of the other circle is given by

$$ax^2 + by^2 - 1 + \{(b - a)/l^2\} (lx + n)^2 = 0$$

$$\Rightarrow \quad ax^2 + by^2 - 1 + (b - a) (x - e)^2 = 0, \qquad ...(6)$$

where $\alpha = -n/l$.

From the equation (2), the equations of the corresponding chords of contact of the circles (5) and (6) are

$$my + n = 0 \quad \text{and} \quad lx + n = 0$$

These are clearly at right angles to each other and intersect at the point $(-n/l, -n/m)$ *i.e.*, (e, α).

Now subtracting (6) from (5), we get

$$(x - e)^2 + (y - \alpha)^2 = 0.$$

This is the equation of point circle with centre at (e, α) and coaxal with the two circles given by (5) and (6). Hence the point of intersection (e, α) of the chords of contact is a point circle *i.e.,* a limiting point.

Example 4:

Prove that any two parabolas which have common focus and their axes in opposite directions intersect at right angles.

Solution:

Take the common focus as origin and the axis of one parabola as the positive direction of the x-axis. Then the equations of the two parabolas are respectively given by

$$y^2 = 4a\,(x + a) \qquad ...(1)$$

and $$y^2 = 4a'\,(x - a') \qquad ...(2)$$

Let (x_1, y_1) be a point of intersection of the parabolas (1) and (2). Then (x_1, y_1) lies on both (1) and (2) so that

$$y_1^2 = 4a\,(x_1 + a) \qquad ...(3)$$

and $$y_1^2 = 4a'\,(x_1 - a') \qquad ...(4)$$

Multiplying (3) by a' and (4) by a and adding, we get

$$y_1^2\,(a' + a) = 4aa'\,(a + a') \text{ i.e., } y_1^2 = 4aa'. \qquad ...(5)$$

The equation of the tangent at the point (x_1, y_1) on the parabola (1) is

$$yy_1 = 2a\,(x + x_1) + 4a^2$$

Its gradient $m_1 = 2a/y_1$.

Similarly the gradient m_2 of the tangent at the point (x_1, y_1) on the parabola (2) is given by

$$m^2 = -\,2a'/y_1.$$

Now $$m_1m_2 = (2a/y_1)\,(-2a'/y_1) = -\,(4aa'/y_1^2)$$

$$= -y_1^2/y_1^2 = -\,1, \quad \text{using (5).}$$

Hence the two parabolas (1) and (2) intersect at right angles.

Example 5:

Two conics have double contact. Prove that the polar of the centre of one conic with respect to other is parallel to the common chord.

Solution:

Let the equation of a conic with centre at (0, 0) be

$$ax^2 + 2hxy + by^2 = 1 \qquad ...(1)$$

and let the equation of a chord of the conic (1) be given by

$$lx + my + n = 0 \qquad ...(2)$$

The equation of a conic having double contact with the conic (1) at the points of intersection of (1) and (2) is given by

$$(ax^2 + 2hxy + by^2 - 1) + \lambda(lx + my + n)^2 = 0 \qquad ...(3)$$

Clearly the conics (1) and (3) have the chord (2) as their common chord.

Now the equation of the polar of the centre (0, 0) of the conic (1) with respect to the conic (3) is given by

$$a.0.x + h(0.y + 0.x) + b.0.y - 1 + \lambda\,[l^2.0.x + 2lm\,(0.x + 0.y) + m^2.0.y + ln\,(0 + x) + mn\,(0 + y) + n^2] = 0$$

$\Rightarrow \quad -1 + \lambda\,\{lnx + mny + n^2\} = 0$

$\Rightarrow \quad lnx + mny + n^2 = 1/\lambda \qquad ...(4)$

$\Rightarrow \quad lx + my = (1/\lambda n) - n$

Clearly the equation (4) of the polar is parallel to the common chord (2). This was to be proved.

Example 6:

Find the equation of the circle which has contact of the second order with the conic

$$ax^2 + 2bxy + cy^2 + 2dx = 0 \text{ at the origin.}$$

Solution:

The equation of the given conic is

$$ax^2 + 2bxy + cy^2 + 2dx = 0. \qquad ...(1)$$

The equation of the tangent to the conic (1) at the origin is obtained by equating to zero the lowest degree terms and is $x = 0$. Also the equation of any straight line passing through the origin is given by $y - mx = 0$.

Hence the general equation of the conics having contact of the second order with the given conic (1) at the origin is given by

$$ax^2 + 2bxy + cy^2 + 2dx - \lambda x\,(y - mx) = 0. \qquad ...(2)$$

If the conic (2) represents a circle, we have

the coefficient of $xy = 0$ *i.e.,* $2b - \lambda = 0$ *i.e.,* $\lambda = 2b$

and the coefficient of x2 = the coefficient of y2

i.e., $a + \lambda m = c$ *i.e.,* $m = (c - a)/\lambda = (c - a)/2b$.

Putting these values of λ and m in (2), the required equation of the circle is given by

$$cx^2 + cy^2 + 2dx = 0.$$

Example 7:

Prove that the equation to the circle having double contact with the ellipse $x^2/a^2 + y^2/b^2 = 1$ at the ends of a latus rectum is

$$x^2 + y^2 + 2ae^3x = a^2(1 - e^2 - e^4).$$

Solution:

The equation of the ellipse is

$$x^2/a^2 + y^2/b^2 = 1. \qquad ...(1)$$

The equation of a latus rectum of the ellipse (1) is

$$x = ae/b \qquad ...(2)$$

The equation of any conic having double contact with the conic (1) at the points of intersection of (1) and (2) is given by

$$\left(\frac{x^2}{a^2} + \frac{y^2}{b^2} - 1\right) + \lambda (x - ae)^2 = 0$$

$$\Rightarrow \quad x^2\left(\frac{1}{a^2} + \lambda\right) + \frac{y^2}{a^2} - 2ae\lambda x - 1 + \lambda a^2e^2 = 0. \qquad ...(3)$$

The equation (3) with represent a circle if

the coefficient of x^2 = the coefficient of y^2

i.e., if $(1/a^2) + \lambda = 1/b^2$,

$$\Rightarrow \quad \lambda = 1/b^2 - 1/a^2 = (a^2 - b^2)/(a^2b^2)$$

$$\Rightarrow \quad \lambda = a^2e^2/\{a^4(1 - e^2)\} = e^2/\{a^2(1 - e^2)\}.$$

Putting this value of λ in (3), the equation of the required circle is given by

$$x^2 + y^2 - 2aexb^2 . \frac{e^2}{a^2(1 - e^2)} - b^2 + \frac{b^2a^2e^2.e^2}{a^2(1 - e^2)} = 0$$

$$\Rightarrow \quad x^2 + y^2 - \frac{2ae^3x.a^2(1 - e^2)}{a^2(1 - e^2)} - a^2(1 - e^2) + \frac{a^2e^2a^2(1 - e^2)}{a^2(1 - e^2)} = 0$$

$$x^2 + y^2 - 2ae^3x = a^2(1 - e^2 - e^4).$$

Example 8:

Pairs of tangents are drawn to the conic $\alpha x^2 + \beta y^2 = 1$ so as to be always parallel to conjugate diameters of the conic $ax^2 + 2hxy + by^2 = 1$;

Show that the locus of their points of intersection is the conic

$$ax^2 + 2hxy + by^2 = a/\alpha + b/\beta.$$

Solution:

Let (x_1, y_1) be the point of intersection of tangents to the conic $\alpha x^2 + \beta y^2 = 1$. The equation of such pair of tangents given by

$$SS_1 = T^2$$

$$\Rightarrow \quad (\alpha x^2 + \beta y^2 - 1)(\alpha x_1^2 + \beta y_1^2 - 1) = (\alpha x x_1 + \beta y y_1 - 1)^2$$

$$\Rightarrow \quad \alpha(\beta y_1^2 - 1)x^2 - 2abx_1y_1xy + \beta(\alpha x_1^2 - 1)y^2 + \text{other terms} = 0. \quad ...(1)$$

Let the lines given by (1) be parallel to the lines given by

$$Ax^2 + 2Hxy + By^2 = 0. \quad ...(2)$$

We know that the lines given by (2) will be parallel to the conjugate diameter of the conic $ax^2 + 2hxy + by^2 = 1$, if

$$aB - 2h\,H + bA = 0 \quad ...(3)$$

Hence the lines given by (1) are parallel tot he conjugate diameters of the conic $ax^2 + 2hxy + by^2 = 1$, if

$$\alpha\beta(\alpha x_1^2 - 1) + 2h\,a\beta\,x_1y_1 + b\alpha(\beta y_1^2 - 1) = 0 \quad \text{(using (3))}$$

$$\Rightarrow \quad \alpha\beta(ax_1^2 + by_1^2 + 2hx_1y_1) = a\beta + b\alpha.$$

Dividing by $\alpha\beta$ and generalising (x_1, y_1), the locus of (x_1, y_1) is given by

$$ax^2 + 2hxy + by^2 = a/\alpha + b/\beta.$$

Example 9:

Find the equations of tangents to the conic

$$x^2 + 4xy + 3y^2 - 5x - 6y + 3 = 0,$$

which are parallel to the straight line

$$x + 4y = 0.$$

Solution:

The equation of the conic is

$$x^2 + 4xy + 3y^2 - 5x - 6y + 3 = 0. \quad ...(1)$$

Let the equation to any straight line parallel to $x + 4y = 0$ be ...(2)

$$x + 4y + \lambda = 0$$

The ordinates of the points of intersection of (1) and (2) are given by

$$\{-(4y + \lambda)\}^2 + 4 -(4y \quad \lambda)\,y + 3y^2 + 5(4y + \lambda) - 6y + 3 = 0$$

$$\Rightarrow \quad 16y^2 + 8y\lambda + \lambda^2 - 16y^2 - 4\lambda y + 3y^2 + 20y + 5\lambda - 6y + 3 = 0$$

$$\Rightarrow \quad 3y^2 + 2y(2\lambda + 7) + (\lambda^2 + 5\lambda + 3) = 0. \quad ...(3)$$

The line (2) will be a tangent to the conic (1), if the value of y from (3) are coincident. Therefore we have

$$4(2\lambda + 7)^2 - 4.3(\lambda^2 + 5\lambda + 3) = 0 \text{ (using } B^2 - 4AC = 0)$$

$$\Rightarrow \quad \lambda^2 + 13\lambda + 40 = 0, \quad \Rightarrow \quad (\lambda + 5)(\lambda + 8) = 0$$

$$\Rightarrow \quad \lambda = -5, -8.$$

Putting these values of λ in the equation (2), the required equation of tangents to (1) are

$$x + 4y - 5 = 0$$

and

$$x + 4y - 8 = 1.$$

Example 10:

Find the locus of eh middle points of he chords of the conic

$$ax^2 + 2hxy + by^2 + 2gx + 2fy + c = 0,$$

whose extremities are equidistant from the point (h, k).

Solution:

The equation of the given conic is given by

$$F(x, y) \equiv ax^2 + 2hxy + by^2 + 2gx + 2fy + c = 0. \quad ...(1)$$

Let the middle point of any chord PQ be $M(x_1, y_1)$, so that its equation given by $T = S_1$ is

$$(ax_1 + hy_1 + g)x + (hx_1 + by_1 + f)y + (gx_1 + fy_1 + c) = F(x_1, y_1). \quad ...(2)$$

It is given that the extremities P and Q of the chord PQ are equidistant from the given point R (h, k). Hence PQR is an isosceles triangle with PQ as base. Hence the line joining R (h, k) to the middle point $M(x_1, y_1)$ of PQ is perpendicular to PQ.

$$\therefore \quad \text{slope of (2)} \times \text{slope of } MR = -1$$

$$\Rightarrow \quad -\frac{ax_1 + hy_1 + g}{hx_1 + by_1 + f} \times \frac{y_1 - k}{x_1 - h} = -1$$

$\therefore$ the locus of M (x_1, y_1) is

$$(ax + hy + g)(y - k) - (hx + by + f)(x - h) = 0.$$

Example 11:

A conic has a double contact with the parabola $y^2 = 4ax$. If the chord of contact passes through the vertex, and the conic passes through the focus of the parabola, prove that the locus of the centre of the conic is the parabola $y^2 = a(2x - a)$.

Solution:

The equation of the given parabola is

$$y^2 = 4ax. \qquad ...(1)$$

The equation of a chord of (1) through the vertex (0, 0) of (1) is

$$y = mx. \qquad ...(2)$$

The equation of any conic having double contact with the parabola (1) at the extremities of the chord (2) is given by

$$y^2 = 4ax + \lambda (y - mx)^2 = 0 \qquad ...(3)$$

The conic (3) will pass through the focus (a, 0) of the parabola (1) if we have

$$-4a^2 + \lambda (-ma)^2 = 0.$$

$$\Rightarrow \qquad \lambda = 4/m^2.$$

Putting this value of λ in (3), the equation of the conic becomes

$$m^2 (y^2 - 4ax) + 4 (y - mx)^2 = 0 \qquad ...(4)$$

Now it is required to find the locus of the centre of the conic (4) for varying values of m.

Let $F(x, y) = m^2 (y^2 - 4ax) + 4 (y - mx)^2$.

Then $\dfrac{\partial F}{\partial x} = -4am^2 - 8m(y - mx)$, $\dfrac{\partial F}{\partial y} = 2m^2y + 8(y - mx)$.

The co-ordinates of the centre of the conic (4) are given by

$$\frac{\partial F}{\partial x} = 0, \text{ and } \frac{\partial F}{\partial y} = 0.$$

i.e., they are given by

$$-4am^2 - 8m(y - mx) = 0. \qquad ...(5)$$

and

$$2m^2y + 8(y - mx) = 0. \qquad ...(6)$$

From (5), we have

$$am + 2y = 2mx = 0,$$

$\Rightarrow$ $m = 2y/(2x - a)$. ...(7)

Dividing the equation (5) by m and then adding to the equation (6), we get

$$-4am + 2m^2y = 0,$$

$\Rightarrow$ $m = 2a/y$. ...(8)

Eliminating m between (7) and (8), the required locus is given by

$$2y/(2x - a) = 2a/y.$$

$\Rightarrow$ $y^2 = a\,(2x - a)$

Example 12:

Find the equation of the circle of curvature at the point (at2, 2at) on the parabola $y^2 = 4ax$.

Solution:

The circle of curvature at the point $(at^2, 2at)$ on the parabola $y^2 = 4ax$ is a circle having a contact of the second order with the parabola at the point $(at^2, 2at)$.

The equation of the parabola is

$$y^2 = 4ax. \quad ...(1)$$

The equation of the tangent at the point $(at^2, 2at)$ on (1) is

$$y.2at = 2a\,(x + at^2)$$

$\Rightarrow$ $y - x/t - at = 0$. ...(2)

We know that the common chords of a circle and a conic are equally inclined to the axes. Therefore the equation of the common chord of the circle of curvature and the parabola drawn through the point $(at^2, 2at)$ is given by

$$y - 2at = (-1/t)\,(x - at^2)$$

$\Rightarrow$ $y + x/t - 3at = 0$. ...(3)

Now the general equation of a conic having contact of the second order with the parabola (1) at the point $(at^2, 2at)$ is given by

$$y^2 - 4ax + \lambda\,(y - x/t - at)\,(y + x/t - 3at) = 0 \quad ...(4)$$

The equation (4) will represent a circle if

the coefficient of x^2 = the coefficient of y^2

i.e., $-\lambda/t^2 = 1 + \lambda$, *i.e.,* $\lambda = t^2\,(1 + t^2)$.

Putting this value of λ in (4), the required equation of the circle of curvature is given by

$$y^2 - 4ax - \{t^2/(1 + t^2)\}\,(y - x/t - at)\,(y + x/t - 3at) = 0$$

$$(1 + t^2)(y^2 - 4ax) - (ty - x - at^2) + (ty + x - 3at^2) = 0$$

$$\Rightarrow \quad (1 + t^2)(y^2 - 4ax) - (t^2y^2 - x^2 - 3at^3y + 3axt^2 - at^3y - at^2x + 3a^2t^4) = 0$$

$$\Rightarrow \quad x^2 + y^2 - 2ax(2 + 3t^2) + 4at^3y - 3a^2t^4 = 0.$$

Example 13:

Show that the confocal hyperbola through the point on the ellipse $x^2/a^2 + y^2/b^2 = 1$ whose eccentric angle is α has for its equation $x^2/\cos^2 \alpha = y^2/\sin^2 \alpha = a^2 = b^2$.

Solution:

The equation of the ellipse is

$$x^2/a^2 + y^2/b^2 = 1. \qquad ...(1)$$

The equation of a conic confocal to the ellipse (1) is

$$x^2/(a^2 + \lambda) + y^2/(b^2 + \lambda) = 1. \qquad ...(2)$$

The coordinates of the point on the ellipse (1) whose eccentric angle is α are $(a \cos \alpha, b \sin \alpha)$. If the equation (2) passes through the point $(a \cos \alpha, b \sin \alpha)$, we get

$$a^2 \cos^2 \alpha/(a^2 + \lambda) + b^2 \sin^2 \alpha/(b^2 + \lambda) = 1$$

$$\Rightarrow \quad (b^2 + \lambda) a^2 \cos^2 \alpha + (a^2 + \lambda) b^2 \sin^2 \alpha = (a^2 + \lambda)(b^2 + \lambda)$$

$$\Rightarrow \quad \lambda^2 + \lambda \{(1 - \cos^2 \alpha) a^2 + (1 - \sin^2 \alpha) b^2\} + a^2b^2 - a^2b^2 (\cos^2 \alpha + \sin^2 \alpha) = 0$$

$$\Rightarrow \quad \lambda^2 + \lambda (a^2 \sin^2 \alpha + b^2 \cos^2 \alpha) = 0.$$

$$\therefore \quad \lambda = 0 \text{ or } \lambda = -(a^2 \sin^2 \alpha + b^2 \cos^2 \alpha).$$

But $\lambda = 0$ reduces the equation (2) to the equation (1).

Hence $\lambda = 0$ is neglected.

Now putting $\lambda = -(a^2 \sin^2 \alpha + b^2 \cos^2 \alpha)$ in (2), the required equation of the conic confocal to (1) is given by

$$x^2/\{a^2 - (a^2 \sin^2 \alpha + b^2 \cos^2 \alpha)\} + y^2/\{b^2 - (a^2 \sin 2 \alpha + b^2 \cos^2 \alpha)\} = 1$$

$$\Rightarrow \quad x^2/\{(a^2 - b^2) \cos^2 \alpha\} + y^2/\{-(a^2 - b^2) \sin^2 \alpha) = 1$$

$$\Rightarrow \quad x^2/\cos^2 \alpha - y^2/\sin^2 \alpha = (a^2 - b^2).$$

This is the equation of a hyperbola confocal to the ellipse (1).

Example 14:

If the confocals through (x_1, y_1) to the ellipse $x^2/a^2 + y^2/b^2 = 1$ are

$$\frac{x^2}{a^2+\lambda_1}+\frac{y^2}{b^2+\lambda_1}=1$$

and $$\frac{x^2}{a^2+\lambda_2}+\frac{y^2}{b^2+\lambda_2}=1$$

show that

(i) $x_1^2/a^2 + y_1^2/b^2 - 1 = -\lambda_1\lambda_2\ (a^2b^2)$.

(ii) $x_1^2 + y_1^2 - b^2 = \lambda_1\lambda_2$.

Solution:

The equation of the given ellipse is

$$x^2/a^2 + y^2/b^2 = 1. \qquad ...(1)$$

Let the equation of any confocal to the ellipse (1) be

$$x^2/(a^2+\lambda) + y^2/(b^2+\lambda) = 1.$$

If it passes through the point (x_1, y_1), we get

$$x_1^2/(a^2+\lambda) + y_1^2/(b^2+\lambda) = 1$$

$$\Rightarrow \quad (b^2+\lambda)\,x_1^2 + (a^2+\lambda)\,y_1^2 = (a^2+\lambda)\,(b^2+\lambda) = 0. \qquad ...(2)$$

$$\lambda^2 + \lambda(a^2 + b^2 - x_1^2 - y_1^2) + (a^2b^2 - b^2x_1^2 - a^2y_1^2) = 0.$$

The equation (2) is a quadratic equation in λ and according to the question the two values of λ are given to be λ_1 and λ_2.

We have

$$\lambda_1 + \lambda_2 = -(a^2 + b^2 - x_1^2 - y_1^2).$$

$$\therefore \quad x_1^2 + y_1^2 - a^2 - b^2 = \lambda_1 + \lambda_2. \qquad ...(3)$$

Also $$\lambda_1.\lambda_2 = (a^2b^2 - b^2x_1^2 - a^2y_1^2),$$

so that $$x_1^2/a^2 + y_1^2/b^2 - 1 = \lambda_1\lambda_2/a^2b^2. \qquad ...(3)$$

The relations (3) and (4) are the required results to be proved.

Example 15:

Prove that the two conics

$$ax^2 + 2hxy + by^2 = 1 \quad and \quad a'x^2 + 2h'xy + b'y^2 = 1$$

can be placed so as to be confocal is

$$\frac{(a-b)^2+4h^2}{(ab-h^2)}=\frac{(a'-b')^2+4h'^2}{(a'b'-h'^2)^2}.$$

Solution:

The equation of one conic is

$$ax^2 + 2hxy + by^2 = 1. \quad ...(1)$$

The squares of the semi-axes of the conic (1) are given by

$$(a - 1/r^2)(b - 1/r^2) = h^2$$

$$\Rightarrow \quad (ab - h^2) r^4 - (a + b) r^2 + 1 = 0.$$

Let r_1^2 and r_2^2 be the roots of this equation, then we have

$$r_1^2 + r_2^2 = (a + b)/(ab - h^2) \text{ and } r_1^2 r_2^2 = 1/(ab - h^2).$$

We know that

$$(r_1^2 - r_2^2)^2 = (r_1^2 + r_2^2)^2 - 4r_1^2 r_2^2.$$

$$\therefore \quad r_1^2 - r_2^2 = \sqrt{\{(a + b)^2/(ab - h^2)^2 - 4/(ab - h^2)\}}$$

$$= \sqrt{[\{(a + b)^2 - 4ab + 4h^2\}/(ab - h^2)^2]}$$

$$= \sqrt{\{(a - b)^2 + 4h^2\}}/(ab - h^2). \quad ...(2)$$

The equation of the other conic is

$$a'x^2 + 2h'xy + b'y^2 = 1. \quad ...(3)$$

If R_1^2 and R_2^2 are the squares of the semi-axes of the conic (3), then proceeding similarly as above, we get

$$R_1^2 - R_2^2 = \sqrt{\{(a' - b')^2 + 4h'^2\}}/(a'b' - h'^2). \quad ...(4)$$

Now if the conics (1) and (3) are confocal, we have

$$r_1^2 - r_2^2 = R_1^2 - R_2^2 \; i.e., \; (r_1^2 - r_2^2)^2 = (R_1^2 - R_2^2)^2$$

i.e., $$\{(a + b)^2 + 4h^2\}/(ab - h^2)^2 = \{(a' - b')^2 + 4h'^2\}/(a'b' - h'^2)^2.$$

Example 16:

Prove that the difference of the squares of the perpendicular drawn from the centre on any two parallel tangents to two given confocals is constant.

Solution:

Let the equations of eh two confocals be

$$\frac{x^2}{a^2 + \lambda_1} + \frac{y^2}{b^2 + \lambda_1} = 1 \quad ...(1)$$

and $$\frac{x^2}{a^2 + \lambda_2} + \frac{y^2}{b^2 + \lambda_2} = 1 \quad ...(2)$$

Let the equations of the two parallel tangents to the confocals (1) and (2) be respectively given by

$$x \cos \alpha + y \sin \alpha = p_1 \quad ...(3)$$

and $$x \cos \alpha + y \sin \alpha = p_2 \quad ...(4)$$

where p_1 and p_2 are the lengths of the perpendiculars from the centre (0, 0) to the tangents (3) and (4) respectively.

Again the straight line (3) is a tangent to (1), therefore we have

$$(a^2 + \lambda_1) \cos^2 \alpha + (b^2 + \lambda_1) \sin^2 \alpha = p_1^2. \quad ...(5)$$

Similarly the straight line (4) is a tangent to (2), so we have

$$(a^2 + \lambda_2) \cos^2 \alpha + (b^2 + \lambda_2) \sin^2 \alpha = p_2^2. \quad ...(6)$$

Subtracting (6) from (5), we get

$$p_1^2 - p_2^2 = (\lambda_1 - \lambda_2)(\cos^2 \alpha + \sin^2 \alpha)$$
$$= \lambda_1 - \lambda_2 = \text{constant}.$$

Example 17:

Show that the locus of the pole of a given straight line w.r.t. to a series of confocal conics is a straight line

Solution:

Let the equation of a straight line be

$$lx + my = 1. \quad ...(1)$$

Also let the equation of a series of confocals be

$$x^2/(a^2 + \lambda) + y^2/(b^2 + \lambda) = 1. \quad ...(2)$$

Let (h, k) be the pole of the straight line (1) with respect to the confocals (2). The equation of the polar of (h, k) w.r.t. (2) is given by

$$hx/(a^2 + \lambda) + ky/(b^2 + \lambda) = 1. \quad ...(3)$$

The equations (1) and (3) represent the same straight line. Therefore comparing their coefficients, we have

$$\frac{h}{l(a^2 + \lambda)} = \frac{k}{m(b^2 + \lambda)} = \frac{1}{1}.$$

$$\therefore \quad a^2 + \lambda = h/l$$

and $b^2 + \lambda = k/m$.

Subtracting these relations, we get

$$a^2 - b^2 = h/l - k/m,$$

$$\Rightarrow \quad mh - lk = lm(a^2 + b^2).$$

This is the equation of a straight line which is perpendicular to the given straight line (1).

Example 18:

Prove that the point of intersection of the two perpendicular tangents on to each of two given confocals lines on a circle.

Solution:

Let the equations of the two confocals be

$$\frac{x^2}{(a^2 + \lambda_1)} = \frac{y^2}{(b^2 + \lambda_1)} = 1 \qquad ...(1)$$

and $$\frac{x^2}{(a^2 + \lambda_2)} = \frac{y^2}{(b^2 + \lambda_2)} = 1 \qquad ...(2)$$

Let the equation of a tangent to (1) be

$$x \cos \alpha + y \sin \alpha = p_1, \qquad ...(3)$$

where $$(a^2 + \lambda_1) \cos^2 \alpha + (b^2 + \lambda_1) \sin^2 \alpha = p_1^2. \qquad ...(4)$$

[Remember that the straight line $lx + my = n$ is a tangent to the conic $x^2/a^2 + y^2/b^2 = 1$ if and only if $a^2l^2 + b^2m^2 = n^2$.]

Let a straight line, perpendicular to (3), given by

$$x \sin \alpha - y \cos \alpha = p_2 \qquad ...(5)$$

be a tangent to (2) so that we have

$$(a^2 + \lambda_2) \sin^2 \alpha + (\lambda^2 + \lambda_2) \cos^2 \alpha = p_2^2. \qquad ...(6)$$

It is required to find the locus of the point of intersection of the tangents (3) and (5) and is obtained by eliminating the parameter α between (3) and (5). Hence squaring the equations (3) and (5) and then adding, the equation of the required locus is given by

$$x^2 + y^2 = p_1^2 + p_2^2$$

$$\Rightarrow \quad x^2 + y^2 = a^2 + b^2 + \lambda_1 + \lambda_2. \qquad \text{(using (4) and (6))}$$

This is the equation of a circle.

Example 19:

Prove that the locus of the points of contact of the tangents drawn a given point to a system of confocals is a cubic curve which passes through the given point and through the foci.

If the given point lies on the major axis, prove that the cubic curve reduces to a circle.

Solution:

Let the equation of the system of confocals be given by

$$x^2/(a^2 + \lambda) + y^2/(b^2 + \lambda) = 1, \qquad ...(1)$$

where λ is the parameter.

Let the co-ordinates of the given point be (α, β).

The equation of the chord of contact of the point (α, β) w.r.t. the confocals (1) is given by

$$\alpha x/(a^2 + \lambda) + \beta y/(b^2 + \lambda) = 1. \qquad ...(2)$$

It is required to find the locus of teh points of contact of tangents drawn from (α, β) to (1) for varying values of λ.

The points of contact lie on the intersection of (1) and (2). Multiplying (1) by β and (2) by y and subtracting, we get

$$a^2 + \lambda = x\,(\alpha y + \beta x)/(y - \beta). \qquad ...(3)$$

Similarly $$b^2 + \lambda = y\,(\alpha y + \beta x)/(y - \beta). \qquad ...(4)$$

Subtracting (4) from (3) and thus eliminating λ, we get

$$a^2 - b^2 = \frac{x(\alpha y - \beta x)}{y - \beta} - \frac{y\,(\alpha y - \beta x)}{\alpha - x}$$

$\Rightarrow$ $(a^2 - b^2)(\alpha - x)(y - \beta) = x\,(\alpha y = \beta x)(\alpha - x) - y(\alpha y - \beta x)(y - \beta)$

$\Rightarrow$ $(a^2 - b^2)(\alpha - x)(y - \beta) = (\alpha y = \beta x)(\alpha x - x^2 - y^2 + \beta y),$ ...(5)

which is the required locus.

The equation (5) being of third degree in x and y respresents a cubic curve and passes through the given point (α, β) as well as the foci $(\pm \sqrt{(a^2 - b^2)}, 0)$.

Again if the given point lies on the major axis i.e on $y = 0$, then $\beta = 0$. Putting $\beta = 0$ in the equation (5), the cubic curve (5) reduces to

$$(a^2 - b^2)(\alpha - x)\,y = \alpha y\,(\alpha x - x^2 - y^2)$$

$\Rightarrow$ $y\,[(a^2 - b^2)(\alpha - x) + \alpha\,(x^2 + y^2 - \alpha x)] = 0$

But $y \equiv 0$ because the point of contact cannot lie on the major axis *i.e.*, $y = 0$.

Hence $(a^2 - b^2)(\alpha - x) + \alpha\,(x^2 + y^2 - \alpha x) = 0$

$\Rightarrow$ $\alpha\,(x^2 + y^2) - (a^2 - b^2 + \alpha^2)\,x + \alpha\,(a^2 - b^2) = 0.$

This is the equation of a circle.

Example 20:

If two conics confocal with a given conic

$$x^2/a^2 + y^2/b^2 = 1$$

pass through a given point P and PQ, PR the normals at P to the confocals meet the polar of P with respect tot he given conic in Q and R, then

$$PQ = -\lambda_1/p_1,\ PR = -\lambda_2/p_2,$$

where p1 and p2 are the perpendicular from the centre to the tangents at P to the confocals and λ_1, λ_2 the parameters of the confocals.

Solution:

The equation of the given conic is

$$\frac{x^2}{a^2} + \frac{y^2}{b^2} = 1. \qquad ...(1)$$

The equation of a confocal with parameter λ_1 is given by

$$\frac{x^2}{(a^2 + \lambda_1)} + \frac{y^2}{(b^2 + \lambda_1)} = 1. \qquad ...(2)$$

Let the co-ordinates of the given point P be (h, k).

Since *P* lies on (2), we have

$$\frac{h^2}{(a^2 + \lambda_1)} + \frac{k^2}{(b^2 + \lambda_1)} = 1. \qquad ...(2')$$

The equation of the tangent to (2) at the point P(h, k) on it is

$$\frac{hx}{(a^2 + \lambda_1)} + \frac{ky}{(b^2 + \lambda_1)} = 1. \qquad ...(3)$$

Example 21:

Prove that the points of intersection of the conics

$$ax^2 + 2hxy + by^2 = 1 \text{ and } a'x^2 + 2h'xy + b'y^2 = 1$$

are at the ends of conjugate diameters of the first conic, if

$$ab' + a'b - 2hh' = 2(ab - h^2).$$

Solution:

The given conics are

$$ax^2 + 2hxy + by^2 = 1. \qquad ...(1)$$

and $\quad a'x^2 + 2h'xy + b'y^2 = 1.$ ...(2)

The centre of the conic (1) is the point (0, 0). Subtracting (2) from (1), we get

$$(a - a')\,x^2 + 2\,(h - h')\,xy + (b - b')\,y^2 = 0. \qquad ...(3)$$

Since the equation (3) is a homogeneous equation of second degree in x and y obtained with the help of (1) and (2), therefore it is the combined equation of the pair of straight lines joining the origin *i.e.,* the centre of the conic (1) to the points of intersection of the conics (1) and (2).

The straight lines given by (3) will be the conjugate diameters of the conic (1) if we have

$$aB = 2hH + bA = 0$$

i.e., $\quad a(b - b') - 2h\,(h - h') + b\,(a - a') = 0$

i.e., $\quad ab' + ba' - 2hh' = 2\,(ab - h^2).$ **Proved.**

Example 22:

Find the equation to the equi-conjugate diameters of the conic

$$ax^2 + 2hxy + by^2 = 1.$$

Solution:

The equation of the conic is

$$ax^2 + 2hxy + by^2 = 1. \qquad ...(1)$$

Here the centre of the conic (1) is (0, 0) and the conjugate diameters are to be of equal length. Let the equation of a circle concentric to (1) be

$$x^2 + y^2 = r^2. \qquad ...(2)$$

In view of the above remark the equi-conjugate diameters will lie along the straight line joining the centre (0, 0) to the points of intersection of (1) and (2). Therefore their combined equation is obtained by making (1) homogeneous with the help of (2). Thus the equation of he pair of these diameters is given by

$$ax^2 + 2hxy + by^2 = (x^2 + y^2)/r^2 \qquad ...(3)$$

$$\Rightarrow \quad (ar^2 - 1)x^2 + 2hr^2\,xy + (br^2 - 1)\,y^2 = 0. \qquad ...(3')$$

The lines given by (3') are conjugate diameters of (1).

$$\therefore \quad a\,(br^2 - 1) - 2h.\,hr^2 + b\,(ar^2 - 1) = 0.$$

$$\Rightarrow \quad 2\,(ab - h^2)\,r^2 = a + b,$$

or $\quad r^2 = \dfrac{a + b}{2(ab - h^2)}.$

Putting this value of r^2 in (3), the required equation of equiconjugate diameters is given by

$$ax^2 + 2hxy + by^2 = 2\,(x^2 + y^2)\,(ab - h^2)/(a + b).$$

Example 23:

Find the condition that the straight line

$$lx + my + n = 0 \qquad \text{...(1)}$$

may touch the conic

$$ax^2 + 2hxy + by^2 + 2gx + 2fy + c = 0. \qquad \text{...(2)}$$

Solution:

The straight line (1) will touch the conic (2) if an only if (1) meets (2) at two coincident points.

From (1), $\quad y = -(lx + n)/m.$

Substituting this value of y in (2), we get

$$ax^2 \; \frac{2hx\,(lx + n)}{m} + b\,\frac{(lx + n)^2}{m} + 2gx - \frac{2f(lx + n)}{m} + c = 0$$

$$\Rightarrow \quad x^2\,(am^2 - 2hlm + bl^2) - 2x\,(hmn - bln - gm^2 - flm) + bn^2 - 2fmn + cm^2 = 0. \qquad \text{...(3)}$$

The equation (3) gives the abscissae of the points of intersection of (1) and (2). The straight line (1) will touch (2) if and only if the roots of the quadratic (3) in x^2 are equal. The condition for this is that

$$(hmn + bln - gm^2 - flm)^2 - (am^2 - 2hlm + bl^2)\,(bn^2 - 2fmn + cm^2) = 0.$$

On simplifying and dividing by m^2, we get the required condition as

$$l^2\,(bc - f^2) + m^2\,(ca - g^2) + n^2\,(ab - h^2) + 2mn\,(gh - af) + 2nl\,(hf - bg) + 2lm\,(fg - ch) = 0.$$

Remark:

Proceeding as in Ex. 7, we can show that the straight line $lx + my + n = 0$ touches the conic $x^2/a + y^2/b = 1$ if and only if

$$al^2 + bm^2 = n^2. \qquad \text{(Remember it)}$$

Example 24:

Show that the locus of the points such that the chords of contact of tangents drawn from them to the conic $ax^2 + 2hxy + by^2 = 1$ subtend a right angle at the centre is the conic

$$(a^2 + h^2)\, x^2 + 2h\,(a + b)\;\; xy + (b^2 + h^2)\, y^2 = a + b$$

Solution:

The equation of the given conic is

$$ax^2 + 2hxy + by^2 = 1 \qquad ...(1)$$

Let the point whose locus is required be (x_1, y_1). the equation of the chord of contact of (x_1, y_1), w.r.t. (1) is given by $T = 0$

$$\Rightarrow \quad axx_1 + h\,(xy_1 + yx_1) + byy_1 - 1 = 0$$

$$\Rightarrow \quad x\,(ax_1 + hy_1) + y\,(by_1 + hx_1) = 1 \qquad ...(2)$$

The equation of the pair of lines joining the centre (0, 0) of (1) to the points of intersection of (1) and (2) is

$$ax^2 + 2hxy + by^2 = \{x\,(ax_1 + hy_1) + y\,(by + hx_1)\}^2$$

$$\Rightarrow \quad ax^2 + 2hxy + by^2 = \{x\,(ax_1 + hy_1) + y\,(by + hx_1)\} = 0. \qquad ...(3)$$

Now (2) will subtend a right angle at the centre (0, 0), if the lines given by (3) are at right angles. The condition for this is that in the equation (3), we should have

the coeff. of x^2 + the coeff. of $y^2 = 0$

$$\Rightarrow \quad a - (ax_1 + hy_1)^2 + b - (by_1 + hx_1)^2 = 0$$

$$\Rightarrow \quad (a^2 + h^2)\, x_1^{\,2} + 2h\,(a + b)\, x_1y_1 + (b^2 + h^2)\, y_1^{\,2} = a + b.$$

$\therefore$ the locus of (x_1, y_1) is

$$(a^2 + h^2)\, x^2 + 2h\,(a + b)\, xy + (b^2 + h^2)\, y_1^{\,2} = a + b.$$

Example 25:

Find the equation of the conic passing through the five points (2, 1), (1, 0), (3, –1), (–1, 0) and (3, 2).

Solution:

We know that the equation of a conic can be found to pass through the five given points. For their we proceed as follows:

The equation of the straight line passing through the first two points (2, 1) and (1, 0) is given by

$$y - 1 = \frac{0 - 1}{1 - 2}\,(x - 2)$$

$$\Rightarrow \quad y - x + 1 = 0. \qquad ...(1)$$

Similarly the equation of the straight line passing through the points (3, 1) and (1, 0) is

$$x + 4y + 1 = 0. \qquad ...(2)$$

The equation of the straight line passing through the points (1, 0) and (3, 1) is

$$2y + x - 1 = 0 \qquad ...(3)$$

and the equation of the straight line passing through the points (2, 1) and (–1, 0) is

$$3y = x = 1 = 0 \qquad ...(4)$$

Hence the equation of any conic passing through the first four of the given points is given by

$$(y - x + 1)(x + 4y + 1) + \lambda (2y + x - 1)$$
$$(3y = x - 1) = 0. \qquad ...(5)$$

If this equation (5) of a conic also passes through the fifth point (3, –2), we have

$$(-2 -3 + 1)(3 - 8 + 1) + \lambda (-4 + 3 - 1)(-6 -3 -1) = 0$$

$\Rightarrow$ $$(-4).(-4) + \lambda (-2).(-10) = 0,$$

$\Rightarrow$ $$\lambda = -4/5.$$

Putting the value of λ in (5), the equation of the required conic is given by

$$(y - x + 1)(x + 4y + 1) - (4/5)$$
$$(2y + x - 1)(3y - x - 1) = 0$$

$\Rightarrow$ $$x^2 + 19xy + 4y^2 = 45y - 1 = 0.$$

Second method: Let the equation of the conic be

$$x^2 + 2hxy + by^2 + 2gx + 2fy + 1 = 0. \qquad ...(1)$$

If (1) passes through teh five points (2, 1) (1, 0) (3, 1), (–1, 0) and (3, –2), we have

$$4a + 4h + b + 4g + 2f + 1 = 0 \qquad ...(2)$$

$$a + 2g + 1 = 0 \qquad ...(3)$$

$$9a - 6h + b + 6g - 2f + 1 = 0 \qquad ...(4)$$

$$a - 2g + 1 = 0 \qquad ...(5)$$

and $$9a - 12h + 4b + 6g - 4f + 1 = 0 \qquad ...(6)$$

Adding the equation (3) and (5), we have

$$2a + 2 = 0 \quad \text{or } a = -1.$$

Putting the value of *a* in (3), we get $g = 0$.

Putting the values of *a* and *g* in (2), (4) and (6), we have

$$4h + b + 2f - 3 = 0 \qquad ...(7)$$

$$-6 + b - 2f - 8 = 0 \qquad (8)$$

and $$-12h + 4b - 4f - 8 = 0 \qquad ...(9)$$

Now solving the equations (7), (8) and (9), we get

$$b = -4,\ 2h = -19,\ 2f = 45.$$

Putting the values of the constants a, b, f, g, h in (1), the required equation of the conic is given by

$$-x^2 - 19xy - 4y^2 + 45y + 1 = 0$$

$$x^2 - 19xy - 4y^2 + 45y + 1 = 0$$

3

Centre of a Conic Section

INTRODUCTON

The coordinates of the centre of the conic (1) can be conveniently found by the use of partial differentiation.

Let $F(x, y) \equiv ax^2 + 2hxy + by^2 + 2gx + 2fy + c = 0$.

We have $\dfrac{\partial F}{\partial x} = 2(ax + hy + g)$, $\dfrac{\partial F}{\partial y} = 2(hx + by + f)$.

From the equations (3) and (4), we see that the *centre of the conic $F(x, y) = 0$ is obtained by solving the equations*

$$ax + hy + g = 0 \quad \text{and} \quad hx + by + f = 0$$

i.e.,

$$\frac{\partial F}{\partial x} = 0 \quad \text{and} \quad \frac{\partial F}{\partial y} = 0.$$

Let (x_1, y_1) be the centre of the conic (1). Then referred to the centre as the origin, the coordinates axes remaining parallel to their old directions, the equation of the conic becomes as

$$ax^2 + 2hxy + by^2 + c_1 = 0, \qquad \text{where } c_1 = gx_1 + fy_1 + c$$

The equation $ax^2 + 2hxy + by^2 + c_1 = 0$ can then be put in the standard form $\quad Ax^2 + 2Hxy + By^2 = 1$.

IF (x_1, y_1) ARE THE COORDINATES OF THE CENTRE OF THE HYPERBOLA

$$F(x, y) \equiv ax^2 + 2hxy + by^2 + 2gx + 2fy + c = 0.$$

then to prove that the equation of the asymptotes is

$$F(x, y) = 2F(x_1, y_1)$$

and the equation of the conjugate hyperbola is

$$F(x, y) = 2F(x_1, y_1).$$

The equations of a hyperbola and its asymptotes differ only by a constant term. So let the eon of the asymptotes of the hyperbola F (x, y) = 0 be

$$F (x,) + \lambda = 0, \qquad ...(1)$$

where λ is a constant.

Since the asymptotes of a hyperbolas pass through its centre, therefore the point (x_1, y_1) must satisfy the equation (1). So we have

$$F(x_1, y_1) + \lambda = 0 \quad \text{or} \quad \lambda = -F(x_1, y_1).$$

Putting this value of λ in (1), the equation of the asymptotes of the given hyperbola is given by

$$F(x, y) - F(x_1, y_1) = 0 \qquad ...(2)$$

$$\Rightarrow \qquad F(x, y) = F(x_1, y_1).$$

We know that the equation of the conjugate hyperbola differs from the equation of the asymptotes by the same constant as the equation of the asymptotes differs from the equation of the hyperbola. So adding the constant term $-F(x_1, y_1)$ to the equation (2), we get the equation of the hyperbola conjugate to the given hyperbola as

$$F(x, y) \; - 2F(x_1, y_1) = 0$$

$$\Rightarrow \qquad F(x, y) \; - 2F(x_1, y_1).$$

ASYMPTOTES

To find the equation of the asymptotes of the central conic

$$ax^2 + 2hxy + by^2 + 2gx + 2fy + c = 0. \qquad ...(1)$$

The equation of a conic and its asymptotes differ only by a constant term. So let the equation of the asymptotes of the conic (1) be given as

$$ax^2 + 2hxy + by^2 + 2gx + 2fy + c + \lambda = 0. \qquad ...(2)$$

where λ is to be so chosen that (2) represents a pair of straight lines.

Now the condition for (2) to represent a pair of straight lines is given as

$$ab\,(c + \lambda) + 2fgh - af^2 - bg^2 - (c + \lambda)\,h^2 = 0$$

[Here the constant term in (2) is $c + \lambda$)]

$$\Rightarrow \qquad (abc + 2fgh - af^2 - bg^2 - ch^2) + \lambda\,(ab - h^2) = 0$$

$$\Rightarrow \qquad \Delta + \lambda\,(ab - h^2) = 0, \text{ or } \lambda = -\Delta/(ab - h^2).$$

Putting this value of l in (2), the equation of the asymptotes of (1) is given as

$$ax^2 + 2hxy + by^2 + 2gx + 2fy + c \;\; \frac{\Delta}{ab - h^2} = 0. \qquad ...(3)$$

NATURE OF CONIC

(a) We can find the nature of the conic

$$ax^2 + 2hxy + by^2 + 2gx + 2fy + c = 0 \quad ...(1)$$

from its second degree terms as shown below.

Let $\Delta \neq 0$, so that the equation (1) does not simply represent a pair of straight lines.

The equation of the pair of straight lines passing through the origin and parallel to the asymptotes of the conic (1) is given by

$$ax^2 + 2hxy + by^2 = 0 \quad ...(2)$$

If $h^2 - ab > 0$, the straight lines (2) are real and so the asymptotes are real. Hence in this case the conic (1) is a hyperbola.

If $h^2 - ab = 0$, the second degree terms in (1) are in a perfect square, and so the conic (1) is a parabola.

If $h^2 - ab < 0$, the straight lines (2) are imaginary and so the asymptotes are imaginary. Hence in this cases the conic (1) is an ellipse.

Thus the general equation of the second degree

$$ax^2 + 2hxy + by^2 + 2gx + 2fy + c = 0$$

represents a hyperbola, a parabola or an ellipse according as

$$h^2 - ab >, =, \text{ or } < 0,$$

unless $\Delta = 0$, *when it represents two straight lines.*

(b) *Criteria for* $ax^2 + 2hxy + by^2 + 2gx + 2fy + c = 0$ *to represent a pair of straight lines, circle, parabola, ellipse or hyperbola.*

The equation $ax^2 + 2hxy + by^2 + 2gx + 2fy + c = 0$ represents

(i) a pair of straight lines if $\Delta = 0$,

(ii) a pair of parallel straight lines if $\Delta = 0$ and $h^2 = ab$.

(iii) a circle if $a = b$ and $h = 0$.

(iv) a parabola if $h^2 - ab$ and $\Delta \neq 0$.

(v) an ellipse if $h^2 < ab$ and $\Delta \neq 0$.

(vi) a hyperbola if $h^2 > ab$ and $\Delta \neq 0$.

(vii) a rectangular hyperbola if $\Delta \neq 0$, $h^2 > ab$ and $a + b = 0$.

ECCENTRICITY, COORDINATES OF THE FOCI AND THE EQUATION OF THE DIRECTORIES OF THE CENTRAL CONIC

$$\mathbf{ax^2 + 2hxy + by^2 + 2gx + 2fy + c = 0}$$

The equation of the given central conic is given by

$$ax^2 + 2hxy + by^2 + 2gx + 2fy + c = 0. \qquad ...(1)$$

Let (x_1, y_1) be the centre of the conic (1). Shifting the region to the centre (x_1, y_1) suppose the equation (1) takes the standard form as

$$Ax^2 + 2Hxy + By^2 = 1. \qquad ...(2)$$

Let r_1^2, r_2^2 be the roots of the quadratic $(A - 1/r^2)(B - 1/r^2) = H^2$ in r^2, and let r_1^2 be the algebraically greater value of r^2 and r_2^2 be the algebraically smaller value of r^2.

In the case of an ellipse, the eccentricity e is given by

$$e^2 = 1 - b^2/a^2$$

and in the case of a hyperbola, the eccentricity e is given by

$$e^2 = 1 + b^2/a^2$$

where in either case the symbols a and b have their standard meanings. If the conic (1) is an ellipse, both r_1^2 and r_2^2 are positive and $r_1^2 = a^2$, $r_2^2 = b^2$.

So the eccentricity e is given by $e^2 = 1 - r_2^2/r_1^2$. If the conic (1) is a hyperbola, r_1^2 is positive and r_2^2 is negative and $r_1^2 = a^2$, $r_2^2 = -b^2$.

So the eccentricity e is given by $e^2 = 1 - r_1^2/r_1^2$. Thus in either case the eccentricity e is given as

$$e^2 = 1 - \frac{r_2^2}{r_1^2}. \qquad ...(3)$$

To find the coordinates of the foci, let θ be the inclination of the major or transverse axis to the x-axis.

Then we have tan θ = the gradient of the line represented by the equation

(6) of 7 $= -\dfrac{(A - 1/r_1^2)}{H}$.

First consider the case when the conic (1) is an ellipse. Let $C(x_1, y_1)$ be the centre of the ellipse (1). The foci are the points on the major axis at a distance er_1 from the centre $C(x_1, y_1)$.

Hence the coordinates of the foci of the ellipse (1) are given by

$$(x_1 + er_1 \cos\theta,\ y_1 + er_1 \sin\theta)$$

and $(x_1 - er_1 \cos\theta,\ y_1 - er_1 \sin\theta)$

where $\quad er_1 = \sqrt{(r_1^2 - r_2^2)}$.

The directories are the lines perpendicular to the major axis at a distance r_1/e *i.e.*, $r_1^2/\sqrt{(r_1^2 - r_2^2)}$, from the centre C. Their equations are,

$$(x - x')\cos\theta + (y - y')\sin\theta = \pm\, r_1^2/\sqrt{(r_1^2 - r_2^2)}.$$

In case the conic (1) is a hyperbola the same formulae hold but in this case the value of r_2^2 is negative.

WORKING RULE TO TRACE AND ELLIPSE OR A HYPERBOLA

Let the equation of the given conic be

$$F(x, y) \equiv ax^2 + 2hxy + by^2 + 2gx + 2fy + c = 0. \quad \ldots(1)$$

Here we are giving the summary of the method to be adopted to trance an ellipse or a hyperbola *i.e.,* when the second terms in the equation (1) do not form a perfect square.

Step I. *Centre.* Find the centre (x_1, y_1) of the conic (1) by solving the equations $\frac{\partial F}{\partial x} = 0$ and $\frac{\partial F}{\partial y} = 0$.

Step II. *New constant term and the equation of the conic referred to the centre as origin.* Transferring the origin to the centre (x_1, y_1), the equation (1) becomes as

$$ax^2 + 2hxy + by^2 + c_1 = 0 \quad \ldots(2)$$

where c_1 = the new constant term = $gx_1 + fy_1 + c$.

In case $c_1 = 0$, the equation (1) will represent a pair of straight lines.

Step III. *Standard form of the equation of the conic.* If a ¹ 0, the equation (2) may be rewritten as

$$-\frac{a}{c_1}x^2 - \frac{2h}{c_1}xy - \frac{b}{c_1}y^2 = 1,$$

$\Rightarrow$ $$Ax^2 + 2Hxy + By^2 = 1, \text{ where } A = a/c_1 \text{ etc.} \quad \ldots(3)$$

Step IV. *Length of the axes of the conic.* The squares of the lengths of the semi axes of the conic are the roots of the equation

$$\left(A - \frac{1}{r^2}\right)\left(B - \frac{1}{r^2}\right) = H^2$$

which is a quadratic in r^2.

If r_1^2 and r_2^2 are its roots and both are +ve, the conic will be an ellipse and if they are of opposite signs, the conic will be a hyperbola.

Step V. *Equation of the axes.* The equation of the major axis (in the case of an ellipse) or that of the transverse axis (in the case of a hyperbola) referred to the centre as origin is given as

$$\left(A - \frac{1}{r_1^2}\right)x + HY = 0. \quad \ldots(4)$$

The equation of the other axis is given as

$$\left(A - \frac{1}{r_2^2}\right) x + HY = 0. \qquad \ldots(5)$$

Referred to the original coordinates axes, the equation of the major or the transverse axis, as the case may be is

$$\left(A - \frac{1}{r_1^2}\right)(x - x_1) + H(y - y_1) = 0. \qquad \ldots(6)$$

The equation of the other axis in this case can also be more conveniently written as

$$\left(A - \frac{1}{r_1^2}\right)(y - y_1) - H(x - x_1) = 0. \qquad \ldots(7)$$

Step VI. *The eccentricity* e (if required) is given as

$$e = \sqrt{\left(1 - \frac{r_2^2}{r_1^2}\right)}.$$

Step VII. *The coordinates of the foci (if required).* If q is the inclination of the major (or transverse) axis to the x-axis, then the coordinates of the foci with respect to the original coordinates axes are given as

$(x_1 + er_1 \cos\theta, y + er_1 \sin\theta)$ and $(x_1 - er_1 \cos\theta, y_1 - er_1 \sin\theta)$.

where $\quad er_1 = \sqrt{(r_1^2 - r_2^2)}$.

Step VIII. *The length of the latus rectum* (if required) is

$$2r_2^2$$

Step IX. *Special points.* Find the points where the gives conic (1) meets the original x and y axes. This will enable us to draw the figure of the conic more accurately.

TRACING OF A PARABOLA

If the general equation of the second degree

$$ax^2 + 2hxy + by^2 + 2gx + 2fy + c = 0.$$

represents a parabola, to trace it and also the find the different results concerning this parabola.

The given general equation of the second degree is

$$ax^2 + 2hxy + by^2 + 2gx + 2fy + c = 0. \qquad \ldots(1)$$

We know that if the equation (1) represents a parabola, the second degree terms form a perfect square. So we can put

$ax^2 + 2hxy + by^2 = (\alpha x + \beta y)^2$ where $\alpha^2 = a$, $\beta^2 = b$ and $\alpha\beta = h$.

The equation (1) then takes the form

$$(\alpha x + \beta y)^2\ 2gx + 2fy + c = 0$$

$$\Rightarrow \quad (\alpha x + \beta y)^2 = -(2gx + 2fy + c)$$

$$\Rightarrow \quad (\alpha x + \beta y + \lambda)^2 = 2x(\lambda\alpha - g) + 2y(\lambda\beta - f) + (\lambda^2 - c),$$

...(2) where λ is an arbitrary constant.

We choose λ such that the lines

$\alpha x + \beta y + \lambda = 0$ and $2x(\lambda\alpha - g) + 2y(\lambda\beta - f) + (\lambda^2 - c) = 0$ are at right angles.

Hence $m_1 \times m_2 = -1$

i.e., $$\left[\frac{-\alpha}{\beta}\right] \cdot -\left\{\frac{(\lambda\alpha - g)}{(\lambda\beta - f)}\right\} = -1$$

$$\Rightarrow \quad \lambda\alpha^2 + \lambda\beta^2 = \alpha g + \beta f, \quad \text{or } \lambda = \frac{\alpha g + \beta f}{\alpha^2 + \beta^2} \qquad \text{...(3)}$$

For this value of λ, the coefficient of 2x in the right hand side of (2).

$$= \frac{\alpha^2 g + \alpha\beta f}{\alpha^2 + \beta^2} - g = \frac{\beta(\alpha f - \beta g)}{\alpha^2 + \beta^2}$$

and the coefficient of 2y

$$= \frac{\alpha\beta g + \beta^2 f}{\alpha^2 + \beta^2} - f = \frac{-\alpha(\alpha f - \beta g)}{\alpha^2 + \beta^2}$$

Therefore the equation (2) becomes.

$$(\alpha x + \beta y + \lambda)^2 = \frac{2(\alpha f - \beta g)}{\alpha^2 + \beta^2}(\beta x - \alpha y) + \lambda^2 - c$$

$$\Rightarrow \quad (\alpha x + \beta y + \lambda)^2 = \frac{2(\alpha f - \beta g)}{\alpha^2 + \beta^2}(\beta x - \alpha y + c')$$

where $$c' = \frac{(\lambda^2 - c)(\alpha^2 + \beta^2)}{2(\alpha f - \beta g)} \qquad \text{...(4)}$$

$$\Rightarrow \quad \left\{\frac{\alpha x + \beta y + \lambda}{\sqrt{(\alpha^2 - \beta^2)}}\right\}^2 = \frac{2(\alpha f - \beta g)}{(\alpha^2 + \beta^2)^{3/2}}\left\{\frac{\beta x - \alpha y + c'}{\sqrt{(\alpha^2 + \beta^2)}}\right\}.$$

...(5)

In the equation (5) we have

$$\frac{\alpha x + \beta y + \lambda}{\sqrt{(\alpha^2 - \beta^2)}} \text{ and } \frac{\beta x - \alpha y + c'}{\sqrt{(\alpha^2 + \beta^2)}}$$

are the perpendicular distances of the point (x, y) on the curve from the mutually perpendicular straight lines $\alpha x + \beta y + \lambda = 0$ and $\beta x - \alpha y + c' = 0$ respectively.

Let us transform the coordinate axes so that the straight lines

$$\alpha x + \beta y + \lambda = 0 \qquad \text{...(6)}$$

and

$$\beta x - \alpha y + c' = 0 \qquad \text{...(7)}$$

become the new axes of x and y respectively.

If (X, Y) are the coordinates of the point (x, y) with respect to these new coordinate axes, the we have

X = the perpendicular distance of the point (x, y) from the new y-axis *i.e.,* the line (7)

$= (\beta x - \alpha y + c')/\sqrt{(\alpha^2 + \beta^2)}$

and Y = the perpendicular distance of the point (x, y) from the new x-axis *i.e.,* the line (6)

$= (\alpha x - \beta y + \chi)/\sqrt{(\alpha^2 + \beta^2)}$.

Therefore the equation (5) transforms into $Y^2 = 4p\,X$, ...(8)

where

$$4p = \frac{2(\alpha f - \beta g)}{(\alpha^2 + \beta^2)^{3/2}} \qquad \text{...(9)}$$

The equation (8) is the equation of the parabola in the standard form. The equation of the axis of the parabola is Y = 0 *i.e.,* $\alpha x + \beta y + \lambda = 0$ *i.e.,* the equation (6) and the equation of the tangent at the vertex is X = 0 *i.e.,* $\beta x - \alpha y + c' = 0$ *i.e.,* the equation (7). The length of the latus rectum of the parabola = 4p, given by (9).

The vertex of the parabola is the point of intersection of the lines (6) and (7).

The equation of the latus rectum of the parabola is X = p,

i.e.,

$$\frac{(\beta x - \alpha y + c')}{\sqrt{(\alpha^2 + \beta^2)}} = \frac{(\alpha f - \beta g)}{2(\alpha^2 + \beta^2)^{3/2}}$$

i.e.,

$$\beta x - \alpha y + c' = \frac{(\alpha f - \beta g)}{2(\alpha^2 + \beta^2)} \qquad \text{...(10)}$$

The focus of the parabola is the point of intersection of the axis of the parabola *i.e.,* the line (6) and the latus rectum of the parabola *i.e.,* the line (10).

The equation of the directrix of the parabola is X = –p

$$i.e., \quad \beta x - \alpha y + c' = -\frac{(\alpha f - \beta g)}{2(\alpha^2 + \beta^2)} \qquad ...(11)$$

The foot of the directrix is the point of intersection of the lines (6) and (11).

SOLVED EXAMPLES

Example 1:

Trace the curve

$$11x^2 + 4xy + 14y^2 - 26x - 32y + 23 = 0.$$

Solution:

The given conic is

$$F(x, y) \equiv 11x^2 + 4xy + 14y^2 = 26x = 32y + 23 = 0.$$

The centre (x_1, y_1) is given by the equations

$$\frac{\partial F}{\partial x} = 22x + 4y = 26 = 0, \quad \frac{\partial F}{\partial y} = 4x + 28y = 32 = 0.$$

Solving, *the centre* C is the point (1, 1).

We have $c_1 = gx_1 + fy_1 + c = (-13).1 + (-16).1 + 23 = -6.$

The equation of the conic referred to the centre (1, 1) as the origin is

$$11x^2 + 4xy + 14^2 - 6 = 0$$

$$\Rightarrow \quad \frac{11}{6}x^2 + \frac{2}{3}xy + \frac{7}{3}y^2 = 1, \text{ reducing to the standard form}$$

$$Ax^2 + 2Hxy + By^2 = 1.$$

The square of the semi-axes are given by

$$\left(A - \frac{1}{r^2}\right)\left(B - \frac{1}{r^2}\right) = H^2, \text{ or } \left(\frac{11}{6} - \frac{1}{r^2}\right)\left(\frac{7}{3} - \frac{1}{r^2}\right) = \left(\frac{1}{3}\right)^2$$

$$\Rightarrow \quad (11r^2 - 6)(7r^2 - 3) = 2r^4,$$

$$\Rightarrow \quad 25r^2 - 25r^2 + 6 = 0, \text{ or } (5r^2 - 2)(5r^2 - 3) = 0.$$

$$\therefore \quad r_1^2 = \frac{3}{5}, r_2^2 = \frac{2}{5}.$$

Since both the values of r^2 are + ve, the curve is an ellipse.

The lengths of major and minor axes are $2\sqrt{\frac{3}{5}}$ and $2\sqrt{\frac{2}{5}}$ respectively.

The equation of the major axis referred to the centre C (1, 1) as the origin is

$$\left(A - \frac{1}{r_1^2}\right)x + Hy = 0, \text{ or } \left(\frac{11}{6} - \frac{5}{3}\right)x + \frac{1}{3}y = 0, \text{ or } x + 2y = 0.$$

Hence the equation of the major axis referred to the old origin is $(x - 1) + 2(y - 1) = 0$, or $x + 2y - 3 = 0$.

It can be seen as in the previous examples that the curve does not cut the coordinate axes.

Example 2:

Find the equation to the hyperbola which has $3x - 4y + 7 = 0$ and $4x + 3y + 1 = 0$ for its asymptotes and passes throught he origin. Trace the curve.

Solution:

Since the equation of the hyperbola differs from the equation of the asymptotes only by a constant term, therefore let the equation of the hyperbola be

$$(3x - 4y + 7)(4x + 3y + 1) + \lambda = 0. \qquad \text{...(1)}$$

It passes through the origin (0, 0) therefore $7 + \lambda = 0$ *i.e.*, $\lambda = -7$.

Hence the required equation of the hyperbola is

$$(3x - 4y + 7)(4x + 3y + 1) = 7. \qquad \text{...(2)}$$

We see that the two straight lines $3x - 4y + 7 = 0$ and $3x + 3y + 1 = 0$ are at right angles and so the hyperbola is a rectangular hyperbola. Let these lines be chosen as the new axes of y and x respectively. Now let the coordinates of any point on the hyperbola referred to the old axes be (x, y) and referred to the new axes be (X, Y). Then we have

X = the length of the perpendicular from (x, y) to the line

$$3x - 4y + 7 = 0$$

$$= \frac{3x - 4y + 7}{\sqrt{(9 + 16)}} = \frac{3x - 4y + 7}{5},$$

and Y = the length of the perpendicular form (x, y) to the line

$$4x + 3y + 1 = 0$$

$$= \frac{4x + 3y + 1}{5}$$

Hence the equation (2) of the hyperbola may be written as

$$XY = 7/25$$

which is the equation of the rectangular hyperbola with new coordinate axes as its asymptotes.

Now referred to the old coordinate axes, the equations of the asymptotes are

$$3x = 4y + 7 = 0 \quad ...(3)$$

and $$4x + 3y + 1 = 0 \quad ...(4)$$

The point of intersection of the asymptotes is the centre of the hyperbola. So solving (3) and (4), the centre of the hyperbola is the point (–1, 1).

Again the transverse and the conjugate axes are the bisectors of the angles between the asymptotes (3) and (4). Hence the equations to the axes of the hyperbola (2) are given by

$$\frac{3x - 4y + 7}{\sqrt{(9 + 16)}} = \pm \frac{4x + 3y + 1}{\sqrt{(16 + 9)}}$$

$$\Rightarrow \quad 3x - 4y + 7 = \pm (4x + 3y + 1).$$

Taking the +ive sign, we get the equation of the transverse axis bisecting the angle between (3) and (4) in which the origin lies as

$$x + 7y - 6 = 0$$

Again taking the –ive sign, the equation of the conjugate axis is

$$7x - y + 8 = 0.$$

The lengths of the semi-axes in the case of the rectangular hyperbola xy = c2 are c $\sqrt{2}$, c $\sqrt{2}$.

Hence the lengths of the semi-axes in the present case are $\sqrt{(14)}/5$ and $\sqrt{(14)}/5$.

The shape of the hyperbola is as shown in the figure.

Example 3:

Show that the product of the semi-axes of the ellipse whose equation is

$$x^2 - xy + 2y^2 - 2x - 6y + 7 = 0$$

is $2/\sqrt{7}$ *and that the equation of the axes is*

$$x^2 - y^2 - 2xy + 8y = 0$$

Solution:

The given conic is

$$F(x, y) \equiv x^2 - xy + 2y^2 - 2x - 6y + 7 = 0$$

The centre (x_1, y_1) is given by the equations

$$\frac{\partial F}{\partial x} = 2x - y - 2 = 0, \quad \frac{\partial F}{\partial y} = -x + 4y\ 6 = 0.$$

Solving, *the centre* is the point (2, 2).

We have $c_1 = gx_1 + fy_1 + c = (-1)(2) + (-3)(2) + 7 = -1$.

$\therefore$ the equation of the conic referred to the centre (2, 2) as the origin is

$$x^2 - xy + 2y^2 - 1 = 0$$

The *standard form* of this equation is

$$x^2 - xy + 2y^2 = 1.$$

The squares of the *semi-axes* are given by

$$\left(A - \frac{1}{r^2}\right)\left(B - \frac{1}{r^2}\right) = H^2, \text{ or } \left(1 - \frac{1}{r^2}\right)\left(2 - \frac{1}{r^2}\right) = \left(-\frac{1}{2}\right)^2$$

$$\Rightarrow \quad 4(r^2 - 1)(2r^2 - 1) = r^4$$

$$\Rightarrow \quad 7r^4 - 12r^2 + 4 = 0$$

Let r_1^2 and r_2^2 be the roots of this quadratic in r^2, then

$$r_1^2 + r_2^2 = \frac{12}{7}, \; r_1^2 r_2^2 = \frac{4}{7}.$$

Since $r_1^2 r_2^2$ is positive and $r_1^2 + r_2^2$ is also positive, therefore r_1^2 and r_2^2 are both positive and so the conic is an ellipse. The lengths of the semi axes of the ellipse are r_1 and r_2. We have

$$r_1 r_2 = 2/\sqrt{7}.$$

which proves the first result.

The equations of the axes of the ellipse referred to the centre (2, 2) as origin are

$$\left(A - \frac{1}{r_1^2}\right)x + Hy = 0 \text{ and } \left(A - \frac{1}{r_2^2}\right)x + Hy = 0$$

$\therefore$ the combined equation of the axes referred to the centre (2, 2) as origin is

$$\left\{\left(A - \frac{1}{r_1^2}\right)x + Hy\right\}.\left\{\left(A - \frac{1}{r_2^2}\right)x + Hy\right\} = 0$$

$$\Rightarrow \left(A - \frac{1}{r_1^2}\right)\left(A - \frac{1}{r_2^2}\right)x^2 + Hxy\left\{\left(A - \frac{1}{r_1^2}\right) + \left(A - \frac{1}{r_2^2}\right)\right\} + H^2y^2 = 0$$

$$\Rightarrow \left\{A^2 - \left(\frac{1}{r_1^2} + \frac{1}{r_2^2}\right)A + \frac{1}{r_1^2 r_2^2}\right\}x^2 + Hxy\left\{2A - \left(\frac{1}{r_1^2} + \frac{1}{r_2^2}\right)\right\} + H^2y^2 = 0$$

$$\Rightarrow \left\{A^2 - \frac{r_1^2 + r_2^2}{r_1^2 + r_2^2}A + \frac{1}{r_1^2 r_2^2}\right\}x^2 + Hxy\left\{2A - \frac{r_1^2 + r_2^2}{r_1^2 + r_2^2}\right\} + H^2y^2 = 0$$

$$\Rightarrow \quad \left\{A^2 - \frac{12/7}{4/7}A + \frac{1}{4/7}\right\}x^2 + Hxy\left\{2A - \frac{12/7}{4/7}\right\} + H^2y^2 = 0$$

$$\Rightarrow \quad \left\{1 - 3.1 + \frac{7}{4}\right\}x^2 + \left(-\frac{1}{2}\right).\, xy\,\{2 - 3) + \left(-\frac{1}{2}\right)^2 y^2 = 0$$

$$\Rightarrow \quad x^2 - 2xy - y^2 = 0.$$

Shifting the origin back from the centre (2, 2) to the old origin, the equation of the axes of the conic referred to the old coordinate axes is

$$(x - 2)^2 - 2\,(x - 2)\,(y - 2) - (y - 2)^2 = 0$$

$$\Rightarrow \quad x^2 - 4x + 4 - 2xy + 4x + 4y - 8 = y^2 + 4y - 4 = 0$$

$$\Rightarrow \quad x^2 - y^2 - 2xy + 8y - 8 = 0$$ **Proved.**

Example 4:

Find the coordinates of the centre of the conic

$$11x^2 + 24xy + 9y^2 - 130ax - 60ay + 116a^2 = 0$$

and hence reduce it to the standard form.

Solution:

The general equaton of the conic is given as follows

$$F\,(x, y) \equiv 41x^2 + 24xy + 9y^2 - 130ax - 60ay + 116a^2 = 0.$$

Now we have

$$\frac{\partial F}{\sigma x} = 82x + 24y - 130a = 0 \textit{ i.e., } 41x + 12y - 65a = 0, \qquad \ldots(1)$$

and $$\frac{\partial F}{\partial y} = 24x + 18y - 60a = 0 \textit{ i.e., } 4x + 3y - 10a = 0, \ldots(2)$$

Solving (1) and (2), we get $x = a$, $y = 2a$.

$\therefore$ the centre (x_1, y_1) is the point $(a, 2a)$.

Now $\quad g = \dfrac{1}{2}$ coeff. of $x = -65a$,

$\quad f = \dfrac{1}{2}$ coeff. of $y = -30a$ and $c = 116a^2$.

$\therefore$ the new constant c_1

$$= gx_1 + fy_1 + c = (-65a)\,a + (-30a)\,2a + 116a^2$$

$$= -a^2.$$

Hence the equation of the conic referred to the centre as origin is

$$41x^2 + 24xy + 9y^2 - 9a^2 = 0$$

$$\Rightarrow \qquad \frac{41}{9a^2}x^2 + \frac{24}{9a^2}xy + \frac{1}{a^2}y^2 = 1,$$

which is of the standard form $Ax^2 + 2Hxy + By^2 = 1$. **Proved.**

Example 5:

Obtain the focus, directrix and axis of the parabola

$$\sqrt{(ax)} + \sqrt{(by)} = 1.$$

If its axis passes through a fixed point, show that the locus of the focus is a rectangular hyperbola.

Solution:

The given curve is

$$\sqrt{(ax)} + \sqrt{(by)} = 1.$$

Squaring, $\quad ax + by + 2\sqrt{(abxy)} = 1$

$$\Rightarrow \qquad (ax + by - 1) = -\sqrt{(abxy)}$$

Again squaring,

$$a^2x^2 + b^2y^2 + 1 + 2abxy = 2ax - 2by = 4abxy$$

$$\Rightarrow \qquad (ax - by)^2 = 2ax + 2by - 1$$

$$\Rightarrow \qquad (ax + by + \lambda)^2 = (2a + 2a\lambda)\,x + (2b - 2b\lambda)\,y + \lambda^2 - 1. \qquad ...(1)$$

Now we choose λ such that the lines $ax - by + \lambda = 0$ and

$$2a\,(1 + \lambda)\,x + 2b\,(1 - \lambda)\,y + \lambda^2 - 1 = 0$$

are at right angles

$$\therefore \qquad \frac{a}{b} \times \left\{-\frac{a\,(1 + \lambda)}{b\,(1 - \lambda)}\right\} = -1, \text{ or } a^2 + a^2\lambda = b^2 - b^2\lambda,$$

$$\Rightarrow \qquad \lambda = \frac{b^2 - a^2}{b^2 + a^2}.$$

Putting this value of λ in (1), we have

$$\left(ax - by + \frac{b^2 - a^2}{b^2 + a^2}\right)^2 = \frac{4ab}{b^2 + a^2}\left\{bx + ay - \frac{ab}{b^2 + a^2}\right\}$$

$$\Rightarrow \qquad \left[\frac{ax - by + \{(b^2 - a^2)/(b^2 + a^2)\}}{\sqrt{(a^2 + b^2)}}\right]^2$$

$$= \frac{4ab}{(b^2 + a^2)^{3/2}} \left[\frac{bx + ax - \{ab/(b^2 + a^2)\}}{\sqrt{(a^2 + b^2)}} \right]$$

which is of the standard form $Y^2 = 4pX$.

The length of the latus rectum $= 4p = \dfrac{4ab}{(b^2 + a^2)^{3/2}}$.

The equation of the axis of the parabola is $Y = 0$

$$\Rightarrow \qquad ax - by + \frac{b^2 - a^2}{b^2 + a^2} = 0 \qquad ...(2)$$

The equation of the *tangent at the vertex is X = 0*

$$\Rightarrow \qquad bx + ay + \frac{ab}{b^2 + a^2} = 0 \qquad ...(3)$$

The focus is given by the equations $X = p$, $Y = 0$

i.e.,

$$\frac{bx + ay - \{ab/(b^2 + a^2)\}}{\sqrt{(a^2 + b^2)}} = \frac{ab}{(b^2 + a^2)^{3/2}},\ ax - by + \frac{b^2 - a^2}{b^2 + a^2} = 0$$

i.e., $\qquad bx + ay - \dfrac{2ab}{b^2 + a^2} = 0, \ \ ax - by + \dfrac{b^2 - a^2}{b^2 + a^2} = 0$

Solving these equations, the focus is $\left(\dfrac{a}{b^2 + a^2}, \dfrac{b}{b^2 + a^2} \right)$.

The equation of the directrix is given by $X = -p$

$$\Rightarrow \qquad \frac{bx + ay - \{ab/(b^2 + a^2)\}}{\sqrt{(b^2 + a^2)}} = \frac{-ab}{(b^2 + a^2)^{3/2}}$$

or $\qquad bx + ay = 0.$

To find the locus of the focus. Let (h, k) be a fixed point. If the axis (2) passes through (h, k), we get

$$ah - bk = \frac{a^2 - b^2}{a^2 + b^2}. \qquad ...(4)$$

Let (x, y) be the coordinates of the focus. Then we have

$$x = \frac{a}{a^2 + b^2},\ y = \frac{b}{a^2 + b^2}. \qquad ...(5)$$

$$\therefore \qquad x^2 + y^2 = \frac{a^2 + b^2}{(a^2 + b^2)} = \frac{1}{a^2 + b^2}. \quad \text{or} \quad a^2 + b^2 = \frac{1}{x^2 + y^2}$$

Using this value in (5), we get

$$a = \frac{x}{x^2 + y^2},\ b = \frac{y}{x^2 + y^2}.$$

Also $a^2 - b^2 = \dfrac{x^2 - y^2}{(x^2 + y^2)^2}$.

Now putting the values of a, b $a^2 + b^2$ and $a^2 - b^2$ in (4), we get

$$\frac{hx}{x^2 + y^2} - \frac{ky}{x^2 + y^2} = \frac{x^2 - y^2}{x^2 + y^2}$$

$$\Rightarrow \quad x^2 - y^2 - hx + ky = 0$$

which is the required locus of hte focus and is a rectangular hyperbola because the coeff. of x^2 + the coeff. of $y^2 = 1 - 1 = 0$.

Example 6:

Prove that the principal axes of the conic

$$ax^2 + 2hxy + by^2 + 2gx + 2fy + c = 0$$

are parallel to the lines $(x^2 - y^2)\ h = (a - b)\ xy$.

Solution:

We know that the asymptotes of the given conic are parallel to the lines given by

$$ax^2 + 2hxy + by^2 = 0. \quad ...(1)$$

Equation of the lines bisecting the angles between the lines given by (1) are given by

$$\frac{x^2 - y^2}{a - b} = \frac{xy}{h},$$

or $\quad (x^2 - y^2)\ h = (a - b)\ xy. \quad ...(2)$

Now the principal axes bisect the angles between the asymptotes. Therefore the principal axes are parallel tot he lines given by (2).

Example 7:

If the general equation of second degree namely

$$ax^2 + 2hxy + by^2 + 2gx + 2fy + c = 0$$

represents a rectangular hyperbola, show that its equation referred to its asymptotes as coordinate axes will be $2\ (h^2 - ab)^{3/2}\ xy \pm \Delta = 0$, where Δ has its usual meaning.

Solution:

Shifting the origin to the centre of the conic, the coordinate axes remaining parallel to their original directions, the given equation becomes.

$$ax^2 + 2hxy + by^2 + \Delta/(ab - h^2) = 0. \qquad ...(1)$$

Now let the coordinate axes be rotated so as to coincide with the asymptotes, which are given to be at right angles. We know that the equation of a rectangular hyperbola referred to the asymptotes as the coordinate axes is of the form $xy = k^2$.

Therefore, by this rotation of the coordinate axes, the equation (1) will take the form

$$2Hxy + \Delta/(ab - h^2) = 0. \qquad ...(2)$$

Now we know that if under a rotation of coordinate axes, the expression $ax^2 + 2hxy + by^2$ transforms to

$$a'x^2 + 2h'xy + b'y^2,$$

then $\qquad h^2 - ab = a'b'.$

$$\therefore \qquad h^2 - ab = H^2 - 0 \times 0 = H^2.$$

i.e., $\qquad H = \pm \sqrt{(h^2 - ab)}.$

Putting this value of H in (2), we see that referred to the asymptotes as the coordinate axes, the given conic is

$$2\,(h^2 - ab)^{3/2}\,xy \pm \Delta = 0.$$

One of these equations represents the case in which the hyperbola lies in the first and third quadrants, and the other in which it lies in the second and fourth.

Example 8:

Prove that the locus of the vertices of the rectangular hyperbola $x^2 - y^2 + 2\lambda xy - a^2 = 0$, for different values of λ is the curve whose equation is given by

$$(x^2 + y^2)^2 - a^2\,(x^2 - y^2) = 0.$$

Solution:

The given equation may be written as

$$x^2 + 2\lambda xy - y^2 = a^2 = 0.$$

$$\Rightarrow \qquad \frac{1}{a^2}x^2 + \frac{2\lambda}{a^2}xy - \frac{1}{a^2}y^2 = 1 \qquad ...(1)$$

This is of the standard form $Ax^2 + 2Hxy + By^2 = 1$ whose centre is

(0, 0) and therefore the centre of the conic (1) is also (0, 0).

The squares of the semi-axes are given by

$$\left(A - \frac{1}{r^2}\right)\left(B - \frac{1}{r^2}\right) = H^2, \text{ or } \left(\frac{1}{a^2} - \frac{1}{r^2}\right)\left(-\frac{1}{a^2} - \frac{1}{r^2}\right) = \left(\frac{\lambda}{a^2}\right)^2$$

$\Rightarrow \quad (r^2 - a^2)(-r^2 - a^2) = \lambda^2 r^4,$

or $\quad -r^4 + a^4 = \lambda^2 r^4$

$\Rightarrow \quad (1 + \lambda^2) r^2 = a^4, = ar$

$\quad = \pm a^2/\sqrt{(1 + \lambda^2)}$

$\therefore \quad r_1^2 = a^2/\sqrt{(1 + \lambda^2)}$ and $r_2^2 = -a^2/\sqrt{(1 + \lambda^2)}$.

Now the equation of the transverse axis is

$$\left(A - \frac{1}{r_1^2}\right) x + Hy = 0, \text{ or } \left(\frac{1}{a^2} - \frac{\sqrt{(1 + \lambda^2)}}{a^2}\right) x + \frac{\lambda}{a^2} y = 0$$

$\Rightarrow \quad \{1 - \sqrt{(1 + \lambda^2)}\} x + \lambda y = 0.$ ...(2)

We have to find the locus of the vertices of the hyperbola (1) which are the points of intersection of the hyperbola (1) and the transverse axis (2).

Thus the required locus will be obtained by eliminating λ between (1) and (2).

From equation (1), $\lambda = \dfrac{y^2 - x^2 + a^2}{2xy}$. ...(3)

From equation (2), $x + \lambda y = x\sqrt{(1 + \lambda^2)}$

$\Rightarrow \quad (x + \lambda y)^2 = x^2 (1 + \lambda^2)$

$\Rightarrow \quad x^2 + \lambda^2 y^2 + 2xy\lambda = x^2 + \lambda^2 x^2$

$\Rightarrow \quad \lambda^2 (y^2 - x^2) = -2xy\lambda$

$\Rightarrow \quad \lambda (y^2 - x^2) = -2xy$

$\Rightarrow \quad \dfrac{(y^2 - x^2 + a^2)}{2xy} (y^2 - x^2) = -2xy$, substituting for λ from (3)

$\Rightarrow \quad (y^2 - x^2) + a^2 (y^2 - x^2) = -4x^2y^2$

$\Rightarrow \quad \{(y^2 - x^2)^2 + 4x^2y^2\} - a^2 (x^2 - y^2) = 0$

$\Rightarrow \quad (x^2 + y^2)^2 - a^2 (x^2 - y^2) = 0,$

which is the required locus of the vertices of (1).

Example 9:

Show that the semi axes of the conic

$$ax^2 + 2hxy + by^2 + 2gx + 2fy + c = 0$$

are given by

$$(ab - h^2)^3 r^4 \Delta (a + b) (ab - h^2) r^2 + \Delta^2 = 0.$$

Hence show that the squares of the semi-axes are equal to

$$\frac{-2\Delta}{(ab - h^2) [a + b \pm \sqrt{\{(a - b)^2 + 4h^2\}}]}$$

where Δ has its usual meaning.

Solution:

The equation of the given conic referred to the centre as origin is

$$ax^2 + 2hxy + by^2 + \frac{\Delta}{ab - h^2} = 0$$

where $\Delta = abc + 2fgh - af^2 - bg^2 - ch^2$.

The equation (1) may be written as

$$-\frac{a (ab - h^2)}{\Delta} x^2 - \frac{2h (ab - h^2)}{\Delta} xy - \frac{b (ab - h^2)}{\Delta} y^2 = 1$$

which is of the standard form $Ax^2 + 2Hxy + By^2 = 1$.

The squares of the semi-axes are given by

$$\left(A - \frac{1}{r^2}\right)\left(B - \frac{1}{r^2}\right) = H^2$$

$$\Rightarrow \left\{-\frac{a (ab - h^2)}{\Delta} - \frac{1}{r^2}\right\}\left\{-\frac{b (ab - h^2)}{\Delta} - \frac{1}{r^2}\right\} = \left\{-\frac{h (ab - h^2)}{\Delta}\right\}^2$$

$$\Rightarrow \quad \{ar^2 (ab - h^2) + \Delta\} \{br^2 (ab - h^2) + \Delta\} = h^2 r^4 (ab - h^2)^2$$

$$\Rightarrow \quad (ab - h^2)^3 r^4 + \Delta (a + b) (ab - h^2) r^2 + \Delta^2 = 0 \qquad ...(2)$$

This proves the first result.

Now treating the equation (2) as a quadratic in r^2, we have

$$r^2 = \frac{-\Delta (a + b) (ab - h^2) \pm \sqrt{[\Delta^2 (a + b)^2 (ab - h^2)^2 - 4 (ab - h^2)^3 \Delta^2]}}{2 (ab - h^2)^3}$$

$$= \frac{\Delta (ab - h^2)}{2 (ab - h^2)^3} \left[-(a + b) \pm \sqrt{\{(a + b)^2 - 4 (ab - h^2)\}}\right]$$

$$= \frac{\Delta}{2\,(ab - h^2)^2}\ [-(a + b) \pm \sqrt{\{(a^2 + b^2 + 2ab - 4ab) + 4h^2\}}]$$

$$= \frac{\Delta}{2\,(ab - h^2)^2}\ [-(a + b) \pm \sqrt{\{(a + b)^2 + 4h^2\}}].$$

Now multiplying the Nr. and Dr. by $-(a + b) \mp \sqrt{\{(a - b)^2 + 4h^2\}}$, we have

$$r^2 = \frac{\Delta}{2\,(ab - h^2)^2}\left[\frac{(a + b)^2 - \{(a - b)^2 + 4h^2\}}{-(a + b) \mp \sqrt{\{(a - b)^2 + 4h^2\}}}\right]$$

$$= -\frac{\Delta}{2\,(ab - h^2)^2}\left[\frac{4ab - 4h^2}{(a + b) \pm \sqrt{\{(a - b)^2 + 4h^2\}}}\right]$$

$$= -\frac{-2\Delta}{(ab - h^2)^2\ [a + b \pm \sqrt{\{(a - b)^2 + 4h^2\}}}$$

This proves the second part.

Example 10:

Find the nature of the conic represented by the equation $x^2 + 2xy + y^2 - 2x - 1 = 0$ and hence trace it.

Solution:

The second degree namely $x^2 + 2xy + y^2$ form a perfect square $(x + y)^2$, therefore the given equation represents a parabola. The given equation may be written as

$$(x + y)^2 = 2x + 1$$

$$\Rightarrow \quad (x + y + \lambda)^2 = (2 + 2\lambda)\,x + 2\lambda y + 1 + \lambda^2. \qquad ...(1)$$

Now we choose λ such that the lines $x + y + \lambda = 0$

and $(2 + 2\lambda)\,x + 2\lambda y + \lambda^2 + 1 + 1 = 0$

are at right angles.

$$\therefore \quad (-1) \times \left\{\frac{-(2 + 2\lambda)}{2\lambda}\right\} = -1, \text{ or } \lambda = -\frac{1}{2}.$$

Hence the equation (1) becomes

$$(x + y - \frac{1}{2}) = x - y + \frac{5}{4}$$

$$\Rightarrow \quad X = \frac{x - y + 1/2}{\sqrt{2}} = \frac{1}{\sqrt{2}}\left[\frac{x - y + 5/4}{\sqrt{2}}\right]$$

which is of the standard form $Y^2 = 4pX$

where $X = \dfrac{x - y + 5/4}{\sqrt{2}}$, $Y = \dfrac{x + y - 1/2}{\sqrt{2}}$, $4p = \dfrac{1}{\sqrt{2}}$

The *axis* of the parabola is

$$Y = 0 \text{ i.e., } x + y = \frac{1}{2} = 0. \qquad ...(2)$$

The *tangent at the vertex* is

$$X = 0 \text{ i.e., } x - y + \frac{5}{4} = 0. \qquad ...(3)$$

Solving (2) and (3), the *vertex* A is (–3/8, 7/8).

The latus rectum of the parabola = 4p = $1/\sqrt{2}$.

The *focus* is given by the equations X = p and Y = 0

i.e., $$\frac{x - y + 5/4}{\sqrt{2}} = \frac{1}{4\sqrt{2}} \text{ and } \frac{x + y - 1/2}{\sqrt{2}} = 0$$

i.e., $$4x - 4y + 4 = 0 \text{ and } x + y + \frac{1}{2} = 0.$$

Solving these, *i.e.,* $x - y + 1 = 0$ and $x + y - \dfrac{1}{2} = 0$, the focus S is $\left(-\dfrac{1}{4}, \dfrac{3}{4}\right)$.

Points of intersection of the parabola with the coordinate axes.

The given curve cuts the x-axis in the points given by

$$y = 0 \text{ and } x^2 - 2x - 1 = 0$$

i.e., $$y = 0 \text{ and } x = \frac{2 \pm \sqrt{(4 + 4)}}{2} = 1 \pm \sqrt{2}$$

i.e., in the points $(1 \pm \sqrt{2}, 0)$ and $(1 - \sqrt{2})$.

The given curve cuts the y-axis in the points given by

$$x = 0 \text{ and } y^2 - 1 = 1$$

i.e., $x = 0$ and $y = \pm 1$

i.e., in the points (0, 1) and (0, –1).

The shape of the curve is as given in the previous figure. The curve lies on the origin side of the tangent at the vertex as is clear from the points where the curve cuts the coordinate axes.

Example 11:

Trace the parabola

$$9x^2 + 24xy + 16y^2 - 2x + 14y + 1 = 0.$$

Solution:

The given equation can be written as

$$(3x + 4y)^2 = 2x - 14y - 1$$

$$\Rightarrow \quad (3x + 4y + \lambda)^2 = (2 + 6\lambda)\,x + (-14 + 8\lambda)\,y + \lambda^2 - 1. \qquad ...(1)$$

Now we choose λ such that the lines

$3x + 4y + \lambda = 0$ and $(2 + 6\lambda)\,x + (-14 + 8\lambda)\,y + \lambda^2 - 1 = 0$ are at right angles.

$$\therefore \quad \left[-\frac{3}{4}\right]\cdot\left[-\frac{2 + 6\lambda}{8\lambda - 14}\right] = -1, \text{ or } \quad 6 + 18\lambda = -32\lambda + 56$$

$$\Rightarrow \quad 50\lambda = 50, \quad \text{or} \quad \lambda = 1.$$

Therefore the equation (1) becomes

$$(3x + 4y + 1)^2 = 8x - 6y = 2\,(4x - 3y)$$

$$\Rightarrow \quad \left\{\frac{3x + 4y + 1}{\sqrt{(9 + 16)}}\right\}^2 = \frac{2}{5}\left\{\frac{4x - 3y}{\sqrt{(16 + 9)}}\right\}$$

which is of the standard form $Y^2 = 4pX$,

where $Y = (3x + 4y + 1)/5$, $X = (4x - 3y)/5$, $4p = 2/5$.

The length of the latus rectum $= 4p = 2/5$ and $p = 1/10$.

The equation of the *axis* is $Y = 0$ *i.e.*, $3x + 4y + 1 = 0$ ...(2)

The *tangent at the vertex* is $X = 0$ *i.e.*, $4x - 3y = 0$. ...(3)

Solving (2) and (3), *the vertex* A is $(-3/25, -4/25)$.

The *focus* is given by the equations $X = p$ and $Y = 0$.

i.e., $$\frac{4x - 3y}{5} = \frac{1}{10} \text{ and } \frac{3x + 4y + 1}{5} = 0,$$

i.e., $8x - 6y = 1$ and $3x + 4y + 1 = 0$.

Solving these, the focus S is $(-1/25, -11/50)$.

The points of intersection of the parabola with the coordinate axes. The curve cuts the x-axis in the points

where $y = 0$ *i.e.*, where $9x^2 - 2x + 1 = 0$ which gives imaginary values of x. Therefore the curve does not cut the x-axis.

The curve cuts the y-axis in the points given by

$$x = 0,\ 16y^2 + 14y + 1 = 0$$

i.e., $$x = 0,\ y = \frac{-14 \pm \sqrt{(196 - 64)}}{32} = -0.8 \text{ and } -0.08 \text{ (approximately).}$$

Example 12:

Show that $\dfrac{1}{x+y-a}+\dfrac{1}{x-y+a}+\dfrac{1}{y-x+a}=0$

represents a parabola. Find coordinates of the focus and the equation to the directrix.

Solution:

The given equation may be written as

$$\frac{1}{x+(y-a)}+\frac{1}{x-(y-a)}=\frac{1}{x-y-a}$$

$$\Rightarrow \quad \frac{x-(y-a)+x+(y-a)}{\{x+(y-a)\}\{x-(y-a)\}}=\frac{1}{x-y-a}$$

$$\Rightarrow \quad 2x(x-y-a)=x^2-(y-a)^2$$

$$\Rightarrow \quad x^2+y^2-2xy-2ax-2ay+a^2=0$$

$$\Rightarrow \quad (x-y)^2=2ax+2ay-a)^2. \qquad ...(1)$$

Since the second degree terms form a perfect square, the equation (1) represents a parabola.

The equation (1) may be written as

$$(x-y)^2=2a\,(x+y-\frac{1}{2}a). \qquad ...(2)$$

Clearly the straight lines $x-y=0$ and $x+y-\dfrac{1}{2}a=0$ are at right angles. So the equation (2) may be written as

$$\left[\frac{x-y}{\sqrt{2}}\right]^2=\frac{2a}{\sqrt{2}}\left[\frac{x+y-1/2a}{\sqrt{2}}\right]$$

where $X=\dfrac{x+y-1/2a}{\sqrt{2}},\ Y=\dfrac{x-y}{\sqrt{2}},\ 4p=\dfrac{2a}{\sqrt{2}}=2\sqrt{2}$.

$\therefore$ the length of the latus rectum $=4p=a\sqrt{2}$, and $p=a\sqrt{2}/4$.

The equation of the axis is

$$Y=0,\ \text{i.e.,}\ x-y=0 \qquad ...(3)$$

The tangent at the vertex is

$$X=0,\ \text{i.e.,}\ x+y-\frac{1}{2}a=0 \qquad ...(4)$$

Solving (3) and (4), the vertex A is $(a/4, a/4)$.

The focus is given by the equations X = p and Y = 0

i.e., $$\frac{x+y-1/2a}{\sqrt{2}} = \frac{a\sqrt{2}}{4} \text{ and } \frac{x-y}{\sqrt{2}} = 0$$

i.e., $x + y = a$ and $x = y$.

Solving these, the focus S is (a/2, a/2).

The equation of the directrix is X = –p

$$\Rightarrow \quad \frac{x+y-1/2}{\sqrt{2}} = -\frac{a\sqrt{2}}{4} \quad \text{or} \quad x + y = 0$$

Points of intersection of the curve with the coordinate axes.

The curve meets the x-axis in the points given by

$$y = 0,\ x^2 - 2ax + a^2 = 0$$

i.e., $y = 0,\ (x - a)^2 = 0$

i.e., $y = 0,\ x = a$ and a.

Thus, the curve meets the x-axis in two coincident points (a, 0) and (a, 0) and so the curve touches the x-axis at the point (a, 0).

Similarly, show that the curve touches the y-axis at (0, a).

Example 13:

Trace the parabola

$$16x^2 - 24xy + 9y^2 + 77x - 64y - 64 + 95 = 0.$$

Also find the coordinates of its focus.

Solution:

Since the second degree terms form a perfect square, the given equation represents a parabola.

The given equation can be written as

$$(4x - 3y)^2 = -77x + 64y - 95$$

$$\Rightarrow \quad (4x - 3y + \lambda)^2 = (8\lambda - 77)x + (64 - 6\lambda)y + \lambda^2 - 95. \quad ...(1)$$

We choose λ such that the lines $4x - 3y + \lambda = 0$ and

$$(8\lambda - 77)x + (64 - 6\lambda)y + \lambda^2 - 95 = 0$$

are at right angles.

$$\therefore \quad \frac{4}{3} \times \frac{-(8\lambda - 71)}{(64 - 6\lambda)} = -1, \text{ or } 321 - 308 = 192 - 18\lambda,$$

$$50\lambda = 500, \text{ or}$$

$\therefore$ the equation (1) becomes

$$(4x - 3y + 10)2 = 3x + 4y + 5$$

$$\Rightarrow \quad \left(\frac{4x - 3y + 10}{5}\right)^2 = \frac{1}{5}\left(\frac{3x + 4y + 5}{5}\right)$$

which is of the standard form $Y^2 = 4pX$,

where $$Y = \frac{4x - 3y + 10}{5}, X = \frac{3x + 4y + 5}{5}, 4p = \frac{1}{5}$$

The length of the latus rectum of the parabola $= 4p = \frac{1}{5}$

The equation of the axis of he parabola is $Y = 0$

i.e., $$4x - 3y + 10 = 0. \qquad ...(2)$$

The equation of the tangent at the vertex is $X = 0$

i.e., $$3x + 4y + 5 = 0. \qquad ...(3)$$

Solving (2) and (3), the vertex A is $\left(-\frac{11}{5}, \frac{2}{5}\right)$.

The equation of the latus rectum is $X = p$

i.e., $$\frac{3x + 4y + 5}{5} = \frac{1}{20}$$

i.e., $12x + 16y + 19 = 0.$ ...(4)

The focus is the point of intersection of the straight lines (2) and (4).

Solving (2) and (4), the focus S is $\left(-\frac{217}{100}, \frac{11}{25}\right)$.

The points of intersection of the curve with the coordinate axes.

The x-axis meets the parabola where $y = 0$ *i.e.,* where $16x^2 + 77x + 95 = 0$ which gives imaginary values of x. therefore the parabola does not meet the x-axis.

The y-axis meets the parabola where $x = 0$ *i.e.,* where

$$9y^2 - 64y + 95 = 0$$

$$\Rightarrow \quad 9y^2 - 45y - 19y + 95 = 0$$

$$\Rightarrow \quad 9y(y - 5) - 19(y - 5) = 0$$

$$\Rightarrow \quad (y - 5)(9y - 19) = 0$$

$$\Rightarrow \quad y = 5 \text{ and } \frac{19}{9}.$$

Example 14:

Find the equation of a conic referred to a pair of conjugate diameters at the coordinate axes.

Solution:

The equation of a conic referred tot he centre as origin is

$$Ax^2 + 2Hxy + By^2 = 1. \quad ...(1)$$

A property of conjugate diameters is that out of a pair of conjugate diameters each bisects the chords parallel to the other.

Hence if we choose a pair of conjugate diameters as the co-ordinate axes, then all the chords parallel to the x-axis will be bisected by the y-axis. Thus if (x_1, y_1) is a point on (1), then $(-x_1, y_1)$ will also be a point on (1). Putting these points in (1) successively, we get

$$Ax_1^2 + 2Hx_1y_1 + By_1^2 = 1. \quad ...(2)$$

and $$Ax_1^2 - 2Hx_1y_1 + By_1^2 = 1. \quad ...(3)$$

Subtracting (3) from (2), we get $H = 0$ as $x_1 \neq 0, y_1 \neq 0$.

Putting the value of H in (1), the required equation of a conic referred to a pair of conjugate diameters as the coordinate axes is

$$Ax^2 + By^2 = 1.$$

Example 15:

Trace the curve

$$x^2 - xy - 2y^2 = x - 4y - 2 = 0$$

Solution:

The given conic is

$$F(x, y) \equiv x^2 - xy - 2y^2 - x - 4y - 2 = 0.$$

Since the second degree terms do not form a perfect square, the given equation represents a central conic. The centre (x_1, y_1) is given by the equations

$$\frac{\partial F}{\partial x} = 2x - y - 1 = 0, \frac{\partial F}{\partial y} = -x - 4y - 4 = 0.$$

Solving these, the centre is $(0, -1)$.

We have $c_1 = gx_1 + fy_1 + c = (-\frac{1}{2})(0) + (-2)(-1) - 2 = 0.$

$\therefore$ the equation of the conic referred to the centre $(0. -1)$ as origin is

$$x^2 - xy - 2y^2 = 0 \quad ...(1)$$

which represents a pair of straight lines passing through the new origin.

The equation (1) may be written as $(x + y)(x - 2y) = 0$.

$\therefore$ the two straight lines represented by (1) are

$$x + y = 0 \text{ and } x - 2y = 0.$$

Shifting the origin back from the point (0, –1) tot he old origin, the equations of these lines referred to the old origin are

$$(x - 0) + \{y - (-1)\} = 0 \text{ i.e., } x + y + 1 = 0$$

and $$(x - 0) - 2\{y - (-1)\} = 0 \text{ i.e., } x - 2y - 2 = 0.$$

Example 16:

A and B are two fixed points and P a variable point. The angle PAB is θ and angle PBA is ϕ. Prove that:

(i) *If $a \tan\theta + b \tan\phi = c$, the locus of P is, in general, a hyperbola passing through A and B; but if $c = 0$, it is a straight line perpendicular to AB; and if $a = b$, it is a parabola whose axis is perpendicular to AB.*

(ii) *If $\sin\theta + \mu \sin\phi$, show that the locus of P is a circle.*

Solution:

Suppose the line BA is taken as the x-axis, its middle point as the origin O and a line perpendicular to BA and passing through O as the y-axis. Then the coordinates of A and B may be taken as $(\lambda, 0)$ and $(-\lambda, 0)$ respectively. Let P be any point (x, y).

The slope of PA $= \tan(180° - \theta) = \dfrac{y}{x - \lambda}$.

$$\therefore \tan\theta = -\frac{y}{x-\lambda}, \text{ so that } \sin\theta = \frac{y}{\sqrt{\{(x-\lambda)^2 + y^2\}}}.$$

Also the slope of

$$PB = \tan\phi = \frac{y}{x+\lambda}, \text{ so that } \sin\phi = \frac{y}{\sqrt{\{(x+\lambda)^2 + y^2\}}}$$

(i) Given $a \tan\theta + b \tan\phi = c$.

$$\therefore a\left(\frac{-y}{x-\lambda}\right) + b.\left(\frac{y}{x+\lambda}\right) = c$$

$$\Rightarrow \quad -ay(x+\lambda) + by(x-\lambda) = c(x^2 - \lambda^2)$$

$$\Rightarrow \quad -xy(a-b) - \lambda y(a+b) = cx^2 - c\lambda^2$$

$\Rightarrow \quad cx^2 + (a - b)\,xy + \lambda\,(a + b)\,y - c\lambda^2 = 0.$ ...(1)

The equation (1) represents a conic section, since it is an equation of second degree in x and y.

We have $H^2 - AB = \left(\dfrac{a-b}{2}\right)^2 - 0$ $\quad [\because$ coeff. of $y^2 = 0]$

$$= \left(\frac{a-b}{2}\right)^2, \quad ...(2)$$

which is, in general, positive.

Hence, in general, the equation (1) *i.e.,* the locus of P is a hyperbola passing through A $(\lambda, 0)$ and B $(-\lambda, 0)$.

(a) But if c is zero, the equation (1) gives

$$(a - b)\,xy + \lambda\,(a + b)\,y = 0$$

$$\Rightarrow \quad y\,[(a - b)\,x + \lambda\,(a + b)] = 0$$

or $$x = \frac{\lambda\,(a+b)}{b-a}$$

which is a straight line perpendicular to x-axis *i.e.,* the locus of P is a straight line perpendicular to AB (since here AB is chosen as x-axis).

(b) If a = b, then from (2), we have $H^2 - AB = 0$. therefore the conic (1) represents a parabola. In this case the equation (1) reduces to

$$cx^2 + 2a\lambda y - c\lambda^2 = 0$$

$$\Rightarrow \quad cx^2 = -\,2a\lambda y - c\lambda^2,$$

$$\text{or } x^2 = \frac{-2a\lambda}{c}\left(y - \frac{c\lambda}{2a}\right) \quad ...(3)$$

The equation (3) is the standard equation of a parabola whose axis is given by x = 0 *i.e.,* the y-axis. Hence the axis of the parabola is perpendicular to AB.

(ii) If $\sin\theta = \mu \sin\phi$, we have $\sin^2\theta = \mu^2 \sin^2\phi$

$$\Rightarrow \quad \frac{y^2}{(x-\lambda)^2 + y^2} = \frac{\mu^2 . y^2}{(x-\lambda)^2 + y^2}$$

$$\Rightarrow \quad (x + \lambda)^2 + y^2 = \mu^2\,(x - \lambda)^2 + \mu^2 y^2$$

$$\Rightarrow \quad x^2 + 2\lambda x + \lambda^2 + y^2 = \mu^2 x^2 - 2\lambda\,\mu^2 x + \mu^2\lambda^2 + \mu^2 y^2$$

$$\Rightarrow \quad (x^2 + y^2) - \mu^2\,(x^2 + y^2) + 2\lambda x\,(1 + \mu^2) + \lambda^2\,(1 - \mu^2) = 0$$

$$\Rightarrow \quad (1 - \mu^2)(x^2 + y^2) + 2\lambda(1 + \mu^2)x + \lambda^2(1 - \mu^2) = 0$$

$$\Rightarrow \quad x^2 + y^2 + \frac{2\lambda(1 + \mu^2)}{1 - \mu^2}x + \lambda^2 = 0$$

which is the equation of a circle.

Example 17:

If the general equation of second degree namely

$$ax^2 + 2hxy + by^2 + 2gx + 2fy + c = 0$$

represents a rectangular hyperbola, show that its equation referred to its asymptotes as coordinate axes will be $2(h^2 - ab)^{3/2} xy \pm \Delta = 0$*, where* Δ *has its usual meaning.*

Solution:

Shifting the origin to the centre of the conic, the coordinate axes remaining parallel to their original directions, the given equation becomes.

$$ax^2 + 2hxy + by^2 + \Delta/(ab - h^2) = 0. \qquad ...(1)$$

Now let the coordinate axes be rotated so as to coincide with the asymptotes, which are given to be at right angles. We know that the equation of a rectangular hyperbola referred to the asymptotes as the coordinate axes is of the form $xy = k^2$.

Therefore, by this rotation of the coordinate axes, the equation (1) will take the form

$$2Hxy + \Delta/(ab - h^2) = 0. \qquad ...(2)$$

Now we know that if under a rotation of coordinate axes, the expression $ax^2 + 2hxy + by^2$ transforms to

$$a'x^2 + 2h'xy + b'y^2,$$

then $\quad h^2 - ab = a'b'$.

$$\therefore \quad h^2 - ab = H^2 - 0 \times 0 = H^2.$$

i.e., $\quad H = \pm\sqrt{(h^2 - ab)}$.

Putting this value of H in (2), we see that referred to the asymptotes as the coordinate axes, the given conic is

$$2(h^2 - ab)^{3/2} xy \pm \Delta = 0.$$

One of these equations represents the case in which the hyperbola lies in the first and third quadrants, and the other in which it lies in the second and fourth.

Example 18:

Determine the centre and the lengths of the axes of the curve $x^2 - 5xy + y^2 + 8x = 20y + 15 = 0$. Reduce it to the standard equation.

Also trace the above conic.

Solution:

The given conic is

$$F(x, y) \equiv x^2 - 5xy + y^2 + 8x - 20y + 15 = 0$$

The centre (x_1, y_1) is given by the equations

$$\frac{\partial F}{\partial x} = 2x - 5y + 8 = 0, \quad \frac{\partial F}{\partial y} = -5x + 2y - 20 = 0.$$

Solving these, *the centre* is the point (–4, 0).

We have $c_1 + gx_1 + fy_1 + c = 4(-4) + 0 + 15 = -1$.

The equation of the curve referred to the centre (–4, 0) as origin is

$$x^2 - 5xy + y^2 - 1 = 0$$

$$\Rightarrow \quad x^2 - 5xy + y^2 = 1, \quad \ldots(1)$$

putting in the standard form $Ax^2 + 2Hxy + By^2 = 1$.

The squares of the *semi-axes* are given by

$$\left(A - \frac{1}{r^2}\right)\left(B - \frac{1}{r^2}\right) = H^2, \text{ or } \left(1 \frac{1}{r^2}\right)\left(1 - \frac{1}{r^2}\right) = \left(-\frac{5}{2}\right)^2$$

$$\Rightarrow \quad (r^2 - 1)^2 = \frac{25}{4} r^4, \text{ or } r^2 - 1 = \pm \frac{5}{2} r^2$$

$$\Rightarrow \quad r^2 + \frac{5}{2} r^4 = 1 \text{ and } r^2 - \frac{5}{2} r^2 = 1.$$

$$\therefore \quad r_1^2 = 2/7, \; r_2^2 = -2/3.$$

Since r_1^2 and r_2^2 are of opposite signs, the conic is a hyperbola. The length of transverse and conjugate axes are $2\sqrt{(2/7)}$ and $2\sqrt{(2/3)}$ respectively.

The equation of the *transverse axis* referred to the centre (–4, 0) as origin is

$$\left(A - \frac{1}{r_1^2}\right)x + Hy = 0, \text{ or } \left(1 - \frac{7}{2}\right)y = 0$$

$$\Rightarrow \quad x + y = 0, \text{ or } y = -x, \text{ or } y = \tan 135^\circ \, x.$$

Hence, the transverse axis is inclined at an angle of 135° to the x-axis. Now we shall turn the coordinate axes through an angle of 135°, so that the transverse axis becomes the x-axis and the conjugate axis becomes the y-axis.

So replacing x by (x cos 135° – y sin 135°) and y by (x sin 135° + y cos 135°) in the equation (1), we get

$$\left[\frac{-x}{\sqrt{2}}-\frac{y}{\sqrt{2}}\right]^2-5\left[-\frac{x}{\sqrt{2}}-\frac{y}{\sqrt{2}}\right]\left[\frac{x}{\sqrt{2}}-\frac{y}{\sqrt{2}}\right]+\left[\frac{x}{\sqrt{2}}-\frac{y}{\sqrt{2}}\right]^2=1$$

$$\Rightarrow \quad (x+y)^2+5(x+y)(x-y)+(x-y)^2=2$$

$$\Rightarrow \quad x^2+y^2+2xy+5x^2-5y^2+x^2+y^2-2xy=2$$

$$\Rightarrow \quad 7x^2-3y^2=2$$

$$\Rightarrow \quad \frac{x^2}{2/7}-\frac{y^2}{2/3}=1,$$

which is the equation of the hyperbola in the standard form.

Example 19:

Trace the conic $6x^2 + 5xy - 6y^2 - 4x + 7y + 11 = 0$

Solution:

The given conic is

$$F(x, y) \equiv 6x^2+5xy+6y^2-4x+7y+11=0.$$

The centre (x_1, y_1) is given by the equations

$$\frac{\partial F}{\partial x}=12x+5y-4=0, \quad \frac{\partial F}{\partial y}=5x-12y+7=0.$$

Solving these, *the centre C* is the point $\left(\frac{1}{13}, \frac{8}{13}\right)$

We have $c_1 + gx_1 + fy_1 + c = (-2)\frac{1}{13}+\frac{7}{2}\cdot\frac{8}{13}+1=13.$

The equation of the curve referred to the centre as origin is

$$6x^2+5xy-6y^2+13=0$$

$$\Rightarrow \quad -\frac{6}{13}x^2+\frac{5}{13}xy+.\frac{6}{13}y^2=1, \text{ reducing to the standard form}$$

$$Ax^2+2Hxy+By^2=1$$

The squares of the *semi-axes* are given by

$$\left(A-\frac{1}{r^2}\right)\left(B-\frac{1}{r^2}\right)=H^2, \text{ or } \left(-\frac{6}{13}-\frac{1}{r^2}\right)\left(\frac{6}{13}-\frac{1}{r^2}\right)=\left(-\frac{5}{26}\right)^2$$

$$\Rightarrow \quad -\frac{36}{169}+\frac{1}{r^4}=\frac{25}{169\times 4}, \text{ or } \frac{1}{r^4}\frac{1}{4},$$

or $\quad r^4 = 4, \quad$ or $\quad r^2 = \pm 2.$

$\therefore \quad r_1^{\,2} = 2$ and $r_2^{\,2} = -2.$

Since the two values of r^2 are of opposite signs, therefore the conic is a hyperbola. Also both the values of r^2 are equal in magnitude and so the hyperbola is rectangular. We have, the length of the transverse axis = the length of the conjugate axis = $2\sqrt{2}$.

The equation of the transverse axis referred to the centre as origin is

$$\left(A - \frac{1}{r_1^2}\right) x + Hy = 0, \quad \text{or} \quad \left(-\frac{6}{13} - \frac{1}{2}\right) x - \frac{5}{26} y = 0,$$

$\Rightarrow \quad 5x + y = 0.$

Therefore the equation of the transverse axis referred to the old origin is $5\left(x - \frac{1}{13}\right) + \left(y - \frac{8}{13}\right) = 0,$

$\Rightarrow \quad 5x + y = 1.$

Also the equation of the conjugate axis referred to the old origin is $x - \frac{1}{13} - 5\left(y - \frac{8}{13}\right) = 0,$ or $x - 5y + 3 = 0.$

Points of intersection of the given conic with the coordinate axes. The curve cuts the x-axis in the points where $y = 0$ *i.e.*, where $6x^2 - 4x + 11 = 0$ which gives imaginary values of x. Therefore the hyperbola does not cut the x-axis.

The curve cuts the y-axis in the points where $x = 0$ *i.e.*, where $6y^2 = 7y - 11 = 0$

i.e., where $y = \dfrac{7 \pm \sqrt{(49 + 264)}}{12} = 20, -0.9$ (nearly).

Example 19:

Prove that the lengths of the semi-axes of the conic

$$ax^2 + 2hxy + ay^2 = d \text{ are } \sqrt{\left(\frac{d}{a+h}\right)} \text{ and } \sqrt{\left(\frac{d}{a-h}\right)}$$

respectively and that their equation is $x^2 - y^2 = 0$.

Solution:

The given equation may be written as

$$\frac{a}{d} x^2 + \frac{2h}{d} xy + \frac{a}{d} y^2 = 1. \qquad \text{...(1)}$$

This is of the standard form $Ax^2 + 2Hxy + By^2 = 1$ whose centre is the origin (0, 0).

The squares of the semi-axes are given by

$$\left(A - \frac{1}{r^2}\right)\left(B - \frac{1}{r^2}\right) = H^2, \text{ or } \left(\frac{a}{d} - \frac{1}{r^2}\right)\left(\frac{a}{d} - \frac{1}{r^2}\right) = \left(\frac{h}{d}\right)^2$$

$$\Rightarrow \quad \left(\frac{a}{d} - \frac{1}{r^2}\right)\left(\frac{h}{d}\right)^2 \quad \text{or} \quad \frac{a}{d} - \frac{1}{r^2} = \pm\frac{h}{d}$$

Taking the positive sign, we have

$$\frac{a}{d} - \frac{h}{d} = \frac{1}{r^2}, \text{ or } r^2 = \frac{d}{a-h}$$

Again taking the negative sign, we have

$$\frac{a}{d} + \frac{h}{d} = \frac{1}{r^2}, \text{ or } r^2 = \frac{d}{a+h}$$

Let $r_1 = \sqrt{\left(\frac{d}{a-h}\right)}$ and $r_2 = \sqrt{\left(\frac{d}{a+h}\right)}$. Then r_1 and r_2 are the lengths of the semi-axes of the conic.

The equation of an axis of the conic is

$$\left(A - \frac{1}{r_1^2}\right)x + Hy = 0, \text{ or } \left(\frac{a}{d} - \frac{a-h}{d}\right)x + \frac{h}{d}y = 0,$$

$$\Rightarrow \qquad x + y = 0.$$

The other axis of the conic is a line perpendicular to the line $x = y = 0$ and passing through the centre (0, 0). So its equation is $x - y = 0$.

Hence the combined equation of the axes is

$$(x + y)(x - y) = 0, \text{ i.e., } x^2 - y^2 = 0.$$

Example 20:

Show that the latus rectum of the parabola

$$(a^2 + b^2)(x^2 + y^2) = (bx + ay - ab)^2$$

is of length $2ab/\sqrt{(a^2 + b^2)}$.

Solution:

The given equation may be written as

$$a^2x^2 + a^2y^2 + b^2x^2 + b^2y^2 = b^2x^2 + a^2y^2 + a^2b^2 + 2abxy - 2ab^2x - 2a^2by$$

$$\Rightarrow \quad a^2x^2 + a^2y^2 - 2abxy = -2ab\left(bx + ay - \frac{1}{2}ab\right)$$

$$\Rightarrow \quad (ax - by)^2 = -2ab\left(bx + ay - \frac{1}{2}ab\right), \qquad ...(1)$$

which is a parabola because the second degree terms form a perfect square.

Clearly the lines $ax - by = 0$ and $bx + ay - \frac{1}{2}ab = 0$ are at right angles. Therefore the equation (1) may be written as

$$\left\{\frac{ax - by}{\sqrt{(a^2 + b^2)}}\right\}^2 = \frac{-2ab}{\sqrt{(a^2 + b^2)}}\left\{\frac{bx + ay - 1/2ab}{\sqrt{(a^2 + b^2)}}\right\}$$

which is of the standard form $Y^2 = 4pX$.

The length of the latus rectum of this parabola is $\dfrac{-2ab}{\sqrt{(a^2 + b^2)}}$.

Example 21:

Show that the curve given by the equations

$$x = at^2 + bt + c,\ y = a't^2 + b't + c'$$

is a parabola whose latus rectum is $(ab' - a'b)^2/(a^2 + a'^2)^{3/2}$.

Solution:

Shifting the origin to the point (c, c') the given parametric equations of the curve become

$$x + c = at^2 + bt + c,\ y + c' = a't^2 + b't + c'$$

i.e., $\quad x = at^2 + bt,\ y = a't^2 + b't$

i.e., $\quad at^2 + bt - x = 0, \qquad ...(1)$

and $\quad a't^2 + b't - y = 0, \qquad ...(2)$

Solving (1) and (2) for t_2 and t, we have

$$\frac{t^2}{b'x - by} = \frac{t}{ay - a'x} = \frac{1}{ab' - a'b}$$

Eliminating t, we get

$$\frac{b'x - by}{ab' - a'b} = \frac{(ay - a'x)^2}{(ab' - a'b)^2}$$

$$\Rightarrow \quad (ay - a'x)^2 = (ab' - a'b)(b'x - by)$$

$$\Rightarrow \quad (ay - a'x)^2 = p\,(b'x - by), \text{ where } p = ab' - a'b$$

$$\Rightarrow \quad (ay - a'x + \lambda)^2 = pb'x - pby + 2\lambda\,(ay - a'x) + \lambda^2$$

$$\Rightarrow \quad (ay - a'x + \lambda)^2 = (2a\lambda - pb)\,y + (pb' - 2\lambda a')\,x + \lambda^2. \qquad ...(3)$$

Now we choose λ such that the lines

$ay - a'x + \lambda = 0$ and $(2a\lambda - pb)\,y + (pb' - 2\lambda a')\,x + \lambda^2 = 0$ are at right angles.

$$\therefore \quad \frac{a'}{a} \times \frac{-(pb' - 2\lambda a')}{(2a\lambda - pb)} = -1$$

$$\Rightarrow \quad a(2a\lambda - pb) = a'\,(pb' - 2\lambda a') = 0$$

$$\Rightarrow \quad 2a\lambda\,(a^2 + a'^2) = p\,(ab + a'b')$$

$$\Rightarrow \quad 2\lambda = \frac{p\,(ab + a'b')}{(a^2 + a'b^2)}.$$

$$\therefore \quad 2a\lambda - pb = \frac{ap(ab + a'b')}{(a^2 + a'^2)} - pb = \frac{ap\,(ab + a'b') - pb\,(a^2 + a'^2)}{(a^2 + a'^2)}$$

$$= \frac{ap\,a'b' - pb\,a'^2}{(a^2 + a'^2)} = \frac{pa'\,(ab' - (a'b)}{(a^2 + a'^2)}$$

$$= \frac{a'\,p^2}{(a^2 + a'^2)}. \qquad [\because ab' - a'b = p]$$

Similarly $pb' - 2\lambda a' = \dfrac{ap^2}{(a^2 + a'^2)}$.

Hence the equation (3) can be written as

$$(ay - a'x + \lambda)^2 = \frac{p^2}{(a^2 + a'^2)}\left[a'y + ax + \frac{\lambda^2\,(a^2 + a^2)}{p^2}\right]$$

$$\Rightarrow \quad \left[\frac{ay - a' + \lambda}{\sqrt{(a^2 + a'^2)}}\right]^2$$

$$= \frac{p^2}{(a^2 + a'^2)}.\frac{1}{\sqrt{(a^2 + a'^2)}}\left[\frac{a'y + ax + \{\lambda^2\,(a^2 + a'^2)/p^2\}}{\sqrt{(a^2 + a'^2)}}\right]$$

$$\Rightarrow \quad \left[\frac{ay - a'x + \lambda}{\sqrt{(a^2 + a'^2)}}\right]^2 = \frac{p^2}{(a^2 + a'^2)^{3/2}}\left[\frac{a'y + ax + \{\lambda^2(a^2 + a^2 + a'^2)/p^2\}}{\sqrt{(a^2 + a'^2)}}\right],$$

which is of the standard form $Y^2 = 4q\,X$.

Hence the given curve is a parabola. The length of the latus rectum of this parabola

$$= 4q = \frac{p^2}{(a^2 + a'^2)^{3/2}} = \frac{(ab' - a'b)^2}{(a^2 + a'^2)^{3/2}}.$$

Example 22:

Trace $7x^2 + 52xy - 32y^2 - 170x + 140y = 0$ and find the equation of its asymptotes.

Solution:

The given conic is

$$F(x, y) \circ 7x^2 + 52xy - 32y^2 - 170x + 140 = 0.$$

The centre (x_1, y_1) is given by the equations

$$\frac{\partial F}{\partial x} = 14x + 52y - 170 = 0, \frac{\partial F}{\partial y} = 52x - 64y + 140 = 0$$

Solving these, *the centre C* is the point (1, 3).

We have $c_1 = gx_1 + f_1 + c = (-85).(1) + (70).(3) + 0 = 125.$

The equation of hte conic referred to the centre (1, 3) as origin is

$$7x^2 + 52xy - 32y^2 + 125 = 0$$

$$\Rightarrow \quad -\frac{7}{125}x^2 - \frac{52}{125}xy + \frac{33}{125}y^2 = 1,$$

reducing to the standard form $Ax^2 + 2Hxy + By^2 = 1$.

The squares of the *semi-axes* are given by

$$\left[A - \frac{1}{r^2}\right]\left[B - \frac{1}{r^2}\right] = H^2, \text{ or } \left[-\frac{7}{125} - \frac{1}{r^2}\right]\left[\frac{32}{125} - \frac{1}{r^2}\right] = \left[-\frac{26}{125}\right]^2$$

$$\Rightarrow \quad 36r^2 + 125r^2 - 625 = 0,$$

or $\quad (9r^2 - 25)(4r^2 + 25) = 0.$

$$\therefore \quad r_1^2 = 25/9, r_2^2 = -25/4.$$

Since r_1^2 and r_2^2 are of opposite signs, the curve is a hyperbola. The lengths of the transverse and conjugate axes are 10/3 and 5 respectively.

The equation of the transverse axis referred to the centre as origin is

Example 23:

Trace the curve

$$41x^2 + 24xy + 9y^2 - 130ax - 60ay + 116a^2 = 0.$$

Solution:

We have the centre C as the point (a, 20).

The equation of the conic when reduced to standard form is

$$\frac{41}{9a^2}x^2 + \frac{24}{9a^2}xy + \frac{1}{a^2}y^2 = 1.$$

The squares of the *semi-axes* are given by

$$\left(A - \frac{1}{r^2}\right)\left(B - \frac{1}{r^2}\right) = H^2, \text{ or } \left(\frac{41}{9a^2} - \frac{1}{r^2}\right)\left(\frac{1}{a^2} - \frac{1}{r^2}\right) = \left(\frac{12}{9a^2}\right)^2$$

or $\quad 25r^4 - 50a^2r^2 + 9a^2 = 0,$

or $\quad (5r^2 - 9a^2)(5r^2 - a^2) = 0$

$\therefore \quad r_1^2 = 9a^2/5,\ r_2^2 = a^2/5.$

Since both r_1^2 and r_2^2 are positive, therefore the conic is an ellipse.

Length of the major axis is $6a/\sqrt{5}$, and that of the minor axis is $2a/\sqrt{5}$.

The equation of the major axis referred to the centre as the origin is

$$\left(A - \frac{1}{r^2}\right)x + hy = 0, \text{ or } \left(\frac{41}{9a^2} - \frac{5}{9a^2}\right)x + \frac{12}{9a^2}y = 0,$$

$\Rightarrow \quad 3x + y = 0$

$\therefore$ the equation of the major axis referred to the old coordinate axes is $3(x = a) + (y - 2a) = 0$, or $3x + y - 5a = 0$.

We can easily show that the curve does not cut the coordinate axes.

Example 24:

Trace the curve

$$32x^2 + 52xy - 7y^2 - 64x - 52y - 148 = 0,$$

also find the coordinates of its foci and the equations of its axes.

Solution:

The given conic is

$$F(x, y) \equiv 32x^2 + 52xy - 7y^2 - 64x - 52y - 148 = 0.$$

The centre (x_1, y_1) is given by the equations

$$\frac{\partial F}{\partial x} = 64x + 52y - 64 = 0,$$

$$\frac{\partial F}{\partial y} = 52x - 14y - 52 = 0$$

Solving, *the centre* is the point (1, 0).

We have $c_1 = gx_1 + fy_1 + c = (-32).1 + 0 - 148 = -180$.

The equation of the conic referred tot he centre as origin is

$$32x^2 + 52xy - 7y^2 - 180 = 0.$$

When reduced tot he standard form $Ax^2 + 2Hxy + By^2 = 1$, this equation becomes

$$\frac{32}{180}x^2 + \frac{52}{180}xy - \frac{7y^2}{180} = 1.$$

The squares of the *semi-axes* are given by

$$\left(A - \frac{1}{r^2}\right)\left(B - \frac{1}{r^2}\right) = H^2, \text{ or } \left(\frac{32}{180} - \frac{1}{r^2}\right)\left(-\frac{7}{180} - \frac{1}{r^2}\right) = \left(\frac{26}{180}\right)^2$$

or $\quad (32r^2 - 180)(-7r^2 - 180) = 26 \times 26r^4$

or $\quad 900r^2 + 4500r^2 - 180 \times 180 = 0$

or $\quad r^2 + 5r^2 - 36 = 0, \text{ or } (r^2 + 9)(r^2 - 4) = 0.$

$\therefore \quad r_1^2 = 4, r_2^2 = -9$ so that $r_1 = 2, \sqrt{|r_2^2|} = 3.$

Since r_1^2 and r_2^2 are of *opposite signs,* the given curve is a *hyperbola.* The lengths of the transverse axis and the conjugate axis are 4 and 6 respectively.

The equation of the transverse axis referred to the centre (1, 0) as origin is

$$\left(A - \frac{1}{r^2}\right)x + Hy = 0, \text{ or } \left(\frac{32}{180} - \frac{1}{4}\right)x \frac{26}{180}y = 0$$

$\Rightarrow \quad (32 - 45)x + 26y = 0$, or $x - 2y = 0$.

Shifting the origin back from the centre (1, 0) to the old origin, the equation of the transverse axis referred to the old coordinate axes is

$$(x - 1) - 2(y - 0) = 0,$$

or $\quad x - 2y - 1 = 0. \qquad ...(1)$

The equation of the *conjugate axis,* a line perpendicular to the transverse axis (1) and passing through the centre (1, 0) is

$$2(x - 1) + (y - 0) = 0, \text{ or } 2x + y - 2 = 0. \qquad ...(2)$$

The eccentricity e

$$= \sqrt{\left(1 - \frac{r_2^2}{r_1^2}\right)} = \sqrt{\left(1 + \frac{9}{4}\right)} = \sqrt{\left(\frac{13}{4}\right)} = \frac{\sqrt{(13)}}{2}.$$

If θ is the inclination of the transverse axis (1) to the x-axis, then tan $\theta = \frac{1}{2}$. Therefore sin θ = 1/√5, cos θ = 2√5.

Foci : The coordinates of the foci are given by

$(x_1 + er_1 \cos\theta, y_1 + er \sin\theta)$ and $(x_1 - er_1 \cos\theta, y_1 - er \sin\theta)$

i.e., $$\left(1 + \frac{\sqrt{(13)}}{2}.2.\frac{2}{\sqrt{5}}, 0 + \frac{\sqrt{13}}{2}.2.\frac{1}{\sqrt{5}}\right)$$

and $$\left(1 + \frac{\sqrt{13}}{2}.2.\frac{2}{\sqrt{5}}, 0 - \frac{\sqrt{13}}{2}.2.\frac{1}{\sqrt{5}}\right)$$

i.e., $$\left(1 + 2\,\frac{\sqrt{13}}{5}\right) \text{ and } \left(1 - 2\sqrt{\frac{13}{5}}, -\sqrt{\frac{13}{5}}\right)$$

The points of intersection of the curve with the coordinate axes.

The curve cuts the x-axis in the points where y = 0 *i.e.,* where

$$32x^2 + 64x - 48 = 0,$$

or $$8x^2 - 16x - 37 = 0$$

⇒ $$x = \frac{16 \pm \sqrt{[(16)^2 + 4.8.37]}}{16} = 1 \pm \frac{\sqrt{(90)}}{4}$$

$$= -1.3, 3.3 \text{ (nearly)}.$$

Thus the points are (–1.3, 0), (3.30).

The curve cuts the y-axis in the points where x = 0, *i.e.,* where $7y^2 + 52y + 148 = 0$ which gives imaginary values of y.

Therefore the hyperbola does not cut the y-axis.

The figure may be drawn in the following manner. Draw the original coordinate axes OX and OY at right angles to each other.. Plot the point C (1, 0) which is the centre of the hyperbola. Draw the transverse axis x – 2y – 1 = 0 which can be done by taking two points on it as follows:

x	1	0
y	0	–1/2

Let the transverse axis be AA' of length 4 such that

$$CA = CA' = 2.$$

Now draw a line perpendicular to the transverse axis through the centre C, and mark two points B and B' such that CB = CB' = 3; then BB' is the conjugate axis. Complete the rectangle KLMN, so that its diagonals KCM and LCN are the asymptotes of the curve. To get a better shape also plot the points of intersection (–1.3, 0) and (3.3, 0) of the curve and the x-axis. Hence the shape of the curve is as shown in the figure.

Example 25:

Trace the hyperbola

$$x^2 - 3xy + y^2 + 10x = 10y + 21 = 0$$

and find the equations of its axes and asymptotes.

Solution:

The given conic is

$$F(x, y) \equiv x^2 - 3xy + y^2 + 10x - 10y + 21 = 0.$$

The centre (x_1, y_1) is given by the equations

$$\frac{\partial F}{\partial x} = x - 3y + 10 = 0, \frac{\partial F}{\partial y} = -3x + 2y - 10 = 0.$$

Solving these, *the centre C* is the point (–2, 2).

We have $$c_1 = gx_1 + fy_1 + c$$
$$= 5(-2) + (-5).2 + 21 = 1.$$

The equation of the conic referred to the centre C as origin is

$$x^2 - 3xy + y^2 = 1$$

$\Rightarrow$ $-x^2 + 3xy - y^2 = 1$, reducing to the standard form

$$Ax^2 + 2Hxy + By^2 = 1.$$

The squares of the *semi-axes* are given by

$$\left(A - \frac{1}{r^2}\right)\left(B - \frac{1}{r^2}\right) = H^2, \text{ or } \left(-1 - \frac{1}{r^2}\right)\left(-1 - \frac{1}{r^2}\right) = \left(\frac{3}{2}\right)^2$$

$$\Rightarrow \quad (r^2 + 1)^2 = \frac{9r^4}{4}, \text{ or } r^2 + 1 = \pm\frac{3r^2}{2},$$

$$\Rightarrow \quad r^2 + \frac{3}{2}r^2 = -1 \text{ and } r^2 - \frac{3}{2}r^2 = -1.$$

$$\therefore \quad r_1^2 = 2 \text{ and } r_2^2 = -\frac{2}{5}.$$

Since the values of r^2 are of opposite signs, the given curve is a hyperbola.

The length of the transverse axis $= 2r_1 = 2\sqrt{2}$,

and the length of the conjugate axis $= 2\sqrt{|r_2^2|} = 2\sqrt{\left(\frac{2}{5}\right)}$.

The equation of the transverse axis referred to the centre (–2, 2) as origin is

$$\left(A - \frac{1}{r_1^2}\right)x + Hy = 0, \text{ or } \left(-1 - \frac{1}{2}\right)x + \frac{3}{2}y = 0, \text{ or } x - y = 0$$

i.e., it makes an angle of 45° with both the coordinate axis.

The equation of the *transverse axis* referred tot he old origin is

$$(x + 2) - (y - 2) = 0, \text{ or } x - y + 4 = 0 \qquad ...(1)$$

The equation of the *conjugate axis i.e.*, the line perpendicular to (1) and passing through the centre (–2, 2) is

$$(x + 2) + (y - 2) = 0, \text{ or } x + y = 0.$$

The shape of the curve is as shown in the figure.

Asymptotes. We know that the equation of the asymptotes differs from the equation of the hyperbola only by a constant term. So let $x^2 - 3xy + y^2 + 10x - 10y + \lambda = 0$ be the equation of the asymptotes of the given hyperbola.

It will represent a pair of straight lines of $\Delta = 0$

i.e., if $abc + 2fgh - af^2 - bg^2 - ch^2 = 0$

i.e., if $1.1.\lambda + 2(-5)(5)\left(-\frac{3}{2}\right) - 1.(-5)^2 - 1.(5)^2 - \lambda.\left(-\frac{3}{2}\right)^2 = 0$

i.e., if $\lambda = 20$.

Hence the combined equation of the asymptotes is

$$x^2 - 3xy + y^2 + 10x - 10y + 20 = 0.$$

Solving it as a quadratic in x, we have

$$x^2 - x(3y - 10) + (y^2 - 10y + 20) = 0.$$

$$\Rightarrow \quad x = \frac{(3x - 10) \pm \sqrt{[(3y - 10)^2 - 4.1.(y^2 - 10y + 20)]}}{2}$$

$$\Rightarrow \quad 2x = (3y - 10) \pm \sqrt{[9y^2 - 60y + 100 - 4y^2 + 40y - 80]}$$

$$\Rightarrow \quad 2x = (3y - 10) \pm \sqrt{[5y^2 - 20y + 20]}$$

$$\Rightarrow \quad 2x = (3y - 10) \pm \sqrt{(5)}\sqrt{(y - 2)^2}.$$

Hence $2x = 3y - 10 \pm \sqrt{5}(y - 2)$ are the required asymptotes.

Example 26:

Trace the curve $14x^2 - 4xy + 11y^2 - 44x - 58y + 71 = 0$. Find the coordinates of its foci and the length of its latus rectum

Solution:

The *centre* C as the point (2, 3).

Also the *standard form* of the equation of the given curve is

$$\frac{7}{30}x^2 - \frac{1}{15}xy + \frac{11}{60}y^2 = 1.$$

The squares of the *semi-axes* are given by

$$\left(A - \frac{1}{r^2}\right)\left(B - \frac{1}{r^2}\right) = H^2,$$

$$\Rightarrow \quad \left(\frac{7}{30} - \frac{1}{r^2}\right)\left(\frac{1}{60} - \frac{1}{r^2}\right) = \left(-\frac{1}{30}\right)^2$$

$$\Rightarrow \quad (7r^2 - 30)(11r^2 - 60) = 2r^4$$

$$\Rightarrow \quad r^2 - 10r^2 + 24 = 0$$

$$\text{or} \quad (r^2 - 6)(r^2 - 4) = 0.$$

$\therefore$ $r_1^2 = 6, r_2^2 = 4$. Hence, the curve is an ellipse with lengths of major and minor axes as $2\sqrt{6}$ and 4.

The equation of the major axis referred to the centre (2, 3) as origin is

$$\left(A - \frac{1}{r_1^2}\right)x + Hy = 0 \text{ or } \left(\frac{7}{30} - \frac{1}{6}\right)x + \left(-\frac{1}{30}\right)y = 0,$$

$$\Rightarrow \quad 2x - y = 0.$$

The equation of the *major axis* referred to the old coordinate axes is

$$2(x - 2) - (y - 3) = 0 \quad \text{or} \quad 2x - y - 1 = 0.$$

Its slope $= \tan\theta = 2$; $\therefore$ $\sin\theta = 2/\sqrt{5}$, $\cos\theta = 1/\sqrt{5}$.

Eccentricity. $e = \sqrt{\left(1 - \frac{r_2^2}{r_1^2}\right)} = \left(1 - \frac{4}{6}\right) = \frac{1}{\sqrt{3}}.$

Foci. Coordinates of the foci are

$(x_1 + er_1 \cos\theta, y_1 + er_1 \sin\theta)$ and $(x_1 - er_1 \cos\theta, y_1 - er_1 \sin\theta)$

i.e., are $\left(2 + \frac{1}{\sqrt{3}}.\sqrt{6}.\frac{1}{\sqrt{5}}, 3 + \frac{1}{\sqrt{3}}.\sqrt{6}.\frac{2}{\sqrt{5}}\right)$

and $\left(2 - \frac{1}{\sqrt{3}}.\sqrt{6}.\frac{1}{\sqrt{5}}, 3 - \frac{1}{\sqrt{3}}.\sqrt{6}.\frac{2}{\sqrt{5}}\right)$

i.e., are $\left(2 + \sqrt{\frac{2}{5}}, 3 + 2\sqrt{\frac{2}{5}}\right)$ and $\left(2 - \sqrt{\frac{2}{5}}, 8 - 2\sqrt{\frac{2}{5}}\right)$.

Length of the latus rectum is

$$= \frac{2r_2^2}{r_1} = \frac{2.4}{\sqrt{6}} = \frac{4\sqrt{6}}{3}.$$

The points of intersection of the conic with the coordinate axes. The curve meets the x-axis in points given by y = 0 *i.e.*, by

$14x^2 - 44x + 71 = 0$ which gives imaginary values of x. Hence, the curve does not cut the x-axis.

The curve cuts the y-axis in points given by x = 0 *i.e.*, $11y^2 - 58y + 71 = 0$. Solving this equation, we have

$$y = \frac{58 \pm \sqrt{\{(58)^2 - 44 \times 71\}}}{22}$$

one value of y lies between 1 and 2 and the other lies between 3 and 4.

Hence the curve is a shown in the figure.

Example 27:

Find the centre, lengths and equations of the axes of the conic $4(x = 2y + 3)^2 + 9(2x + y = 1)^2 = 80$. *Also trace the conic.*

Solution:

If we simplify the equation, it can be traced by the method given in the above examples. Here we shall give an alternative method for a particular problem of this type.

We see that the two straight lines

$$x - 2y + 3 = 0 \text{ and } 2x + y - 1 = 0$$

are perpendicular to each other. Let these lines be chosen as the new axes of y and x respectively.

Now let the coordinates of any point on the conic referred to the old coordinate axes be (x, y) and referred to the new coordinate axes be (X, Y). Then we have

X = the length of the perpendicular from (x, y) on the line

$$x - 2y + 3 = 0$$

$$= \frac{x - 2y + 3}{\sqrt{5}} \qquad ...(1)$$

and Y = the length of the perpendicular from (x, y) on the line

$$2x + y - 1 = 0$$

$$= \frac{2x + y - 1}{\sqrt{5}} \qquad ...(2)$$

Hence the given equation of the conic may be written as

$$20 = \left\{\frac{x - 2y + 3}{\sqrt{5}}\right\}^2 + 45\left\{\frac{2x + y - 1}{\sqrt{5}}\right\}^2 = 80$$

$$\Rightarrow \quad 20X^2 + 45Y^2 = 80, \qquad \text{(using (1) and (2))}$$

$$\Rightarrow \quad \frac{X^2}{4} + \frac{Y^2}{16/9} = 1$$

which is the equation of an ellipse whose lengths of major and minor axes are 4 and 8/3 respectively.

Also the equation of the major axis is

Y = 0 *i.e.,* x + y – 1 = 0 and the equation oof the minor axis is

X = 0 *i.e.,* x – 2y + 3 = 0 ....(4)

The point of intersection of (3) and (4) is $\left(-\frac{1}{5}, \frac{7}{5}\right)$ which is the centre of the ellipse is as shown in the figure.

Example 28:

Find the axis, the vertex, the latus rectum, the focus, and the equation of the directrix of the parabola

$$16x^2 - 24xy + 9y^2 - 104x - 127y + 44 = 0.$$

Solution:

The second degree terms namely $16x^2 - 24xy + 9y^2$ form a perfect square, therefore the given equation represents a parabola.

We may write the given equation as

$$(4x - 3y)^2 = 104x + 172y - 44. \qquad ...(1)$$

[On the left hand side write only the second degree terms.]

The lines 4x – 3y = 0 and 104x + 172y – 44 = 0 are not at right angles ($\because m_1m_2 \neq -1$), therefore we introduce a new constant λ in the L.H.S. of

(1) and adjust the R.H.S. accordingly. Thus the equation (1) becomes

$$(4x - 3y + \lambda)^2 = (104 + 8\lambda)\,x + (172 - 6\lambda)\,y + \lambda^2 - 44. \quad ...(2)$$

Now we choose λ such that the lines

$$4x - 3y + \lambda = 0 \quad ...(3)$$

and $$(104 + 8\lambda)\,x + (172 - 6\lambda)\,y + \lambda^2 - 44 = 0 \quad ...(4)$$

are at right angles.

For this, we have

$$\left(\frac{4}{3}\right) - \left\{\frac{104 + 8\lambda}{12 - 6\lambda}\right\} = -1$$

$\Rightarrow$ $\quad -4(104 + 8\lambda) = -3\,(172 - 6\lambda)$, or $50\lambda = 100$, or $\lambda = 2$.

Putting this value of λ in (2), we have

$$(4x - 3y) + 2)^2 = 40\,(3x + 4y - 1)$$

$$\Rightarrow \quad 25 \times \left[\frac{4x - 3y + 2}{\sqrt{\{(4)^2 + (-3)^2\}}}\right]^2 = 5 \times 40 \times \left[\frac{3x + 4y - 1}{\sqrt{\{(3)^2 + (4)^2\}}}\right] \quad \text{(Note)}$$

$$\Rightarrow \quad \left\{\frac{4x - 3y + 2}{5}\right\}^2 = 8\left\{\frac{3x + 4y - 1}{5}\right\}. \quad ...(5)$$

The equation (5) is of the standard form

$$Y^2 = 4pX, \quad ...(6)$$

where $X = \dfrac{3x + 4y - 1}{5}$, $Y = \dfrac{4x - 3y + 2}{5}$, $4p = 8$.

The axis of the parabola is $Y = 0$ *i.e.,* $4x - 3y + 2 = 0$. ...(7)

The tangent at the vertex is $X = 0$ *i.e.,* $3x + 4y - 1 = 0$. ...(8)

The vertex of the parabola is the point of intersection of the lines (7) and (8). Solving (7) and (8), the coordinates of the vertex A are (–1/5, 2/5) or (–0.2, 0.4).

The *length of the latus rectum* is 4p *i.e.,* 8.

Also since $4p = 8$, therefore $p = 2$.

The *equation of the latus rectum* is $X = p$,

i.e., $$\frac{3x + 4y - 1}{5} = 2 \text{ i.e., } 3x + 4y = 11. \quad ...(9)$$

The focus of the parabola is the point of intersection of the axis of the parabola *i.e.,* the line (7) and the latus rectum *i.e.,* the line (9). Therefore solving the equations (7) and (9), the coordinates of the focus are (1, 2).

The equation of the directrix is given by $X = -p$

$$\Rightarrow \quad \frac{3x + 4y - 1}{5} = -2, \text{ or } 3x + 4y + 9 = 0$$

Example 29:

Obtain the focus and directrix of the parabola $\sqrt{x} + \sqrt{y} = \sqrt{a}$. *Also trace it.*

Solution:

The equaiton of parabola is

$$\sqrt{x} + \sqrt{y} = \sqrt{a}.$$

Squaring, $x + y + 2\sqrt{(xy)} = a$

$$\Rightarrow \quad (x + y + a) = -2\sqrt{(xy)}.$$

Again squaring, $(x + y + a)^2 = 4xy$

$$\Rightarrow \quad x^2 + y^2 + 2xy - 2ax - 2ay + a^2 = 4xy$$

$$\Rightarrow \quad x^2 + y^2 - 2xy = 2ax + 2ay - a^2$$

$$\Rightarrow \quad (x - y)^2 = 2ax + 2ay - a^2$$

Now do yourself.

Example 30:

Trace the parabola $4x^2 - 4xy + y^2 - 8x - 6y + 5 = 0$

Solution:

Since the second degree terms form a perfect square $(2x - y)^2$, the given equation represents a parabola. The given equation may be written as

$$(2x - y)^2 = 8x + 6y - 5$$

$$\Rightarrow \quad (2x - y + \lambda)^2 = (8 + 4\lambda) x + (6 - 2\lambda) y + \lambda^2 - 5. \quad ...(1)$$

Now we choose λ such that the lines

$$2x - y + \lambda = 0 \text{ a}$$

nd $(8 + 4\lambda) x + (6 - 2\lambda) y + \lambda^2 - 5 = 0$ are at right angles.

$$\therefore \qquad 2 \times -\left(\frac{8+4\lambda}{6-2\lambda}\right) = -1, \text{ or } 16 + 8\lambda = 6 - 2\lambda.$$

$$\Rightarrow \qquad 10\lambda = -10, \text{ or } \lambda = -1.$$

Hence the equation (1) becomes

$$(2x - y - 1)^2 = 4x + 8y + 8y - 4 = 4(x + 2y - 1)$$

$$\Rightarrow \qquad \left(\frac{2x-y-1}{\sqrt{5}}\right)^2 = \frac{4}{\sqrt{5}}\left(\frac{x+2y-1}{\sqrt{5}}\right)$$

which is of the standard form $Y^2 = 4p\,X$

where $X = \dfrac{x+2y-1}{\sqrt{5}}, Y = \dfrac{2x-y-1}{\sqrt{5}}, 4p = \dfrac{4}{\sqrt{5}}$

$\therefore$ the length of the latus rectum

$= 4p = 4/\sqrt{5}$ and $p = 1/\sqrt{5}$.

The axis is $Y = 0$ *i.e.*, $2x - y - 1 = 0$. ...(2)

The *tangent at the vertex* is

$X = 0$ *i.e.*, $x + 2y - 1 = 0$. ...(3)

Solving (2) and (3), the vertex A is (3/5, 1/5).

The *focus* is given by the equations $X = p$ and $Y = 0$,

i.e., $\dfrac{x+2y-1}{\sqrt{5}} = \dfrac{1}{\sqrt{5}}$ and $\dfrac{2x-y-1}{\sqrt{5}} = 0$

i.e., $x + 2y - 2 = 0$ and $2x - y - 1 = 0$

Solving these, the foucs S is (4/5, 3/5).

Points of intersection of the parabola with the coordinate axes. The curve cuts the x-axis in the points where $y = 0$, *i.e.*, where $4x^2 - 8x + 5 = 0$ which gives imaginary values of x. Thus the curve does not cut the x-axis.

The curve meets the y-axis in the points given by

$x = 0$ and $y^2 - 6y + 5 = 0$

i.e., $x = 0$ and $y = 1, 5$.

The points are (0, 1) and (0, 5).

Hence the shape of the curve is as shown in the previous figure.

Example 31:

Find the coordinates of the centre of the conic

$$11x^2 + 24xy + 9y^2 - 130ax - 60ay + 116a^2 = 0$$

and hence reduce it to the standard form.

Solution:

The general equaton of the conic is given as follows

$$F(x, y) \equiv 41x^2 + 24xy + 9y^2 - 130ax - 60ay + 116a^2 = 0.$$

Now we have

$$\frac{\partial F}{\sigma x} = 82x + 24y - 130a = 0 \text{ i.e., } 41x + 12y - 65a = 0,$$

...(1)

$$\text{and } \frac{\partial F}{\partial y} = 24x + 18y - 60a = 0 \text{ i.e., } 4x + 3y - 10a = 0,$$

...(2)

Solving (1) and (2), we get $x = a$, $y = 2a$.

$\therefore$ the centre (x_1, y_1) is the point $(a, 2a)$.

Now $g = \frac{1}{2}$ coeff. of $x = -65a$,

$f = \frac{1}{2}$ coeff. of $y = -30a$ and $c = 116a^2$.

$\therefore$ the new constant c_1

$$= gx_1 + fy_1 + c = (-65a)\,a + (-30a)\,2a + 116a^2$$
$$= -a^2.$$

Hence the equation of the conic referred to the centre as origin is

$$41x^2 + 24xy + 9y^2 - 9a^2 = 0$$

$$\Rightarrow \quad \frac{41}{9a^2}x^2 + \frac{24}{9a^2}xy + \frac{1}{a^2}y^2 = 1,$$

which is of the standard form $Ax^2 + 2Hxy + By^2 = 1$.

Example 32:

Trace the parabola

$$9x^2 - 24xy + 16y^2 - 18x - 101y + 19 = 0$$

and find the coordinates of its focus.

Solution:

The given equation represents a parabola because the second degree terms form a perfect square $(3x - 4y)^2$. The given equation may be written as

$(3x - 4y)^2 = 18x + 101y = 19.$

Introducing an arbitrary constant λ, the above equation may be written as

$$(3x - 4y + \lambda)^2 = (18x + 6\lambda)\, x + (101 - 8\lambda)\, y + \lambda^2 - 19 \qquad ...(1)$$

Now we choose λ such that the lines

$$3x - 4y + \lambda = 0 \text{ and } (18x + 6\lambda)\, x + (101 - 8\lambda)\, y + \lambda^2 - 19 = 0$$

are at right angles.

$$\therefore \quad \frac{3}{4} \times - \left\{\frac{18 + 6\lambda}{101 - 8\lambda}\right\} = -1, \quad \text{or } 54 + 18l = 404 - 32\lambda$$

$$\Rightarrow \qquad 50\lambda = 350, \text{ or } \lambda = 7.$$

Therefore the equation (1) of the parabola becomes

$$(3x - 4y + 7)^2 = 60x + 45y + 30 = 15\,(4x + 3y + 2)$$

$$\Rightarrow \qquad \left\{\frac{3x - 4y + 7}{\sqrt{(9 + 16)}}\right\}^2 = \frac{15}{5}\left\{\frac{4x + 3y + 2}{\sqrt{16 + 9)}}\right\}$$

which is of the standard form $Y^2 = 4pX$

where $\quad X = \dfrac{4x + 3y + 2}{5}, \; Y = \dfrac{3x - 4y + 7}{5}, \; 4p = 3.$

The axis of the parabola is

$$Y = 0 \text{ i.e., } 3x - 4y + 7 = 0 \qquad ...(2)$$

The tangent at the vertex is

$$X = 0 \text{ i.e., } 4x + 3y + 2 = 0. \qquad ...(3)$$

Vertex. Solving (2) and (3),m the vertex A of the parabola is

$$\left(-\frac{29}{25}, \frac{22}{25}\right).$$

The length of the latus rectum $= 4p^2 = 3$. Also $p = 3/4$.

The *focus* is the point of intersection of the lines $X = p$ and $Y = 0$.

i.e., of the lines $\dfrac{4x + 3y + 2}{5} = \dfrac{3}{4}$ and $\dfrac{3x - 4y + 7}{5} = 0.$

i.e., of the lines $16x + 12y - 7 = 0$ and $3x = 4y + 7 = 0$

Solving these equations, the focus S is $\left(-\dfrac{14}{25}, \dfrac{133}{100}\right).$

The points of intersection of the parabola with the coordinate axes. The given curve cuts the x-axis where $y = 0$, *i.e.,* where $9x^2 - 18x + 19 = 0$ which gives imaginary values of x since

$$(18)^2 - 4 \times 9 \times 19 = -\text{ive}.$$

Hence the curve does not cut the x-axis.

Again the curve meets the y-axis where $x = 0$,

i.e., where $$16y^2 - 101y + 19 = 0$$

or $$y = \frac{101 \pm \sqrt{\{(101)^2 - 4 \times 16 \times 19\}}}{2 \times 16}$$

$$= 0.2 \text{ and } 6.1 \text{ (nearly).}$$

Tracing. Draw the rectangular axes Ox and OY and plot the vertex A whose coordinates are

$$\left(\frac{-29}{25}, \frac{22}{25}\right) \quad i.e., \ (-1.16, 0.88).$$

Draw the straight line $3x - 4y + 7 = 0$ which is the axis of the parabola. This line passes through the vertex A and cuts the y-axis at the point $(0, 7/4)$ *i.e.,* $(0, 1.75)$.

Draw the tangent at the vertex *i.e.,* a line through the vertex A and perpendicular to the axis of the parabola.

Since the parabola does not cut the x-axis and it cuts the y-axis at the points whose y-coordinates are approximately 0.2 and 6.1, therefore the curve lies on the origin side of the tangnet at the vertex.

Now mark the focus S on the axis of the parabola such that $AS = p = 3/4 = 0.75$. Draw a line through S and perpendicular to the axis $3x = 4y + 7 = 0$ and take two points L and L' on it such that $SL = SL' = 2p = 3/2 = 1.5$. Also plot the points where the curve cuts the y-axis.

Now draw the parabola touching the tangent at the vertex at the point A, symmetrical about the axis $3x = 4y + 7 = 0$, and passing through the points L, L' and the points on the y-axis whose approximate y-coordinates are 0.2 and 6.1.

Example 33:

Find the coordinates of the focus and vertex of the parabola

$$x^2 - 4xy + 4y^2 - 12x - 6y - 39 = 0.$$

Solution:

Since the second degree terms form a perfect square, the given equation represents a parabola. The given equation may be written as

$$(x - 2y)^2 = 12x + 6y + 39$$

or $$(x - 2y + \lambda)^2 = (12x + 2\lambda)\, x + (6 - 4\lambda)\, y + \lambda^2 + 39. \quad ...(1)$$

Now we choose l such that the lines $x - 2y + \lambda = 0$

and $$(12x + 2\lambda)\, x + (6 - 4\lambda)\, y + \lambda^2 + + 39 = 0$$

are at right angles.

$$\therefore \quad \frac{1}{2} \times \left(- \frac{12 + 2\lambda}{6 - 4\lambda} \right) = -1,$$

or $12 + 2\lambda = 12 - 8\lambda,$

or $10\lambda = 0, \quad$ or $\quad \lambda = 0.$

Hence the equation (1) becomes

$$(x - 2y)^2 = 12x + 6y + 39 = 6\,(2x + y + 13/2)$$

or $$\left(\frac{x - 2y}{\sqrt{5}} \right)^2 = \frac{6}{\sqrt{5}} \cdot \left(\frac{2x + y + 13/2}{\sqrt{5}} \right)$$

which is the standard from $y^2 = 4px$

where $$X = \frac{2x + y + 13/2}{\sqrt{5}}, \; Y = \frac{x - 2y}{\sqrt{5}}, \; 4p = \frac{6}{\sqrt{5}}.$$

$\therefore$ the length of the latus rectum $= 4p = \dfrac{6}{\sqrt{5}}$ and $p = \dfrac{3}{2\sqrt{5}}.$

The *axis* of the parabola is $Y = 0$

i.e., $$x = 2y = 0. \quad ...(2)$$

The *tangent at the vertex* is $X = 0$

i.e., $$2x + y + 13/2 = 0. \quad ...(3)$$

Solving (2) and (3), the *vertex* A is $(-26/10, - 13/10)$.

The *focus* is given by the equations $X = p$ and $Y = 0$.

i.e., $$\frac{2x + y + 13/2}{\sqrt{5}} = \frac{3}{2\sqrt{5}} \text{ and } \frac{x - 2y}{\sqrt{5}} = 0$$

i.e., $4x + 2y + 10 = 0$ and $x = 2y$.

Solving these, the focus S is $(-2, -1)$.

Points of intersection of the curve with the coordinate axes. The curve cuts the x-axis in the points given by

$$y = 0 \text{ and } x^2 - 12x - 39 = 0$$

i.e., $$y = 0 \text{ and } x = \frac{12 \pm \sqrt{(144 + 156)}}{2}$$

i.e., $$y = 0 \text{ and } x = 14.7, -2.7 \text{ (nearly).}$$

$\therefore$ the points are (14.7, 0), (–2.7, 0).

The curve cuts the y-axis in the points given by

$$x = 0,\ 4y^2 - 6y - 39 = 0$$

i.e., $$x = 0,\ y = \frac{6 \pm \sqrt{(36 + 624)}}{8}$$

i.e., $$x = 0,\ y = \frac{6 \pm 25.7}{8}$$

The points are (0, 4) and (0, – 2.5) nearly.

Hence the shape of the curve is as shown in the figure.

The curve lies on the origin side of the tangent at the vertex.

Example 34:

Find the equation of the hyperbola whose asymptotes are parallel to $2x + 3y = 0$ and $3x + 2y = 0$, whose centre is at (1, 2) and which passes through (5, 3).

Solution:

Let the equations of the asymptotes parallel to the given lines be as

$$2x + 3y = \lambda_1 \text{ and } 3x + 2y = \lambda_2.$$

Since these pass through the centre (1, 2, therefore we have

$$\lambda_2 = 8 \text{ and } \lambda_2 = 7.$$

Thus the equations of the asymptotes are as follows

$$2x + 3y - 8 = 0 \text{ and } 3x + 2y - 7 = 0.$$

Now let the equation of the hyperbola be

$$(2x + 3y - 8)(3x + 2y - 7) = \lambda.$$

If it passes through (5, 3), then $\lambda = 154$.

$\therefore$ the required equation of the hyperbola is given by

$$(2x + 3y - 8)(3x + 2y - 7) = 154$$

Example 35:

Find the equation of the asymptotes of the conic

$$3x^2 - 2xy - 5y^2 + 7x - 9y = 0$$

and find the equation of the conic which has the same asymptotes and which passes through the point (2, 2).

Solution:

The equation of the asymptotes differs from the equation of the conic only by a constant term, therefore let the equation of the asymptotes be given as

$$3x^2 - 2xy - 5y^2 + 7x - 9y + \lambda = 0. \qquad ...(1)$$

where λ is a constant to be determined by the fact that (1) should represent a pair of straight lines.

The equation (1) will represent a pair of straight lines if

$$abc + 2fgh - af^2 - bg^2 - ch^2 = 0$$

$$\Rightarrow\ 3\,(-5)\lambda + 2\left(-\frac{9}{2}\right)\left(\frac{7}{2}\right)(-1) - 3\left(-\frac{9}{2}\right)^2 - (-5)\left(\frac{7}{2}\right)^2 - \lambda\,(-1)^2 = 0$$

$$\Rightarrow\ \lambda = 2$$

∴ the equation of the asymptotes is

$$3x^2 - 2xy - 5y^2 + 7x - 9y + 2 = 0. \qquad ...(2)$$

Now let the equation of a conic having (2) for its asymptotes be

$$3x^2 - 2xy - 5y^2 + 7x - 9y + 2 + \mu = 0. \qquad ...(3)$$

where m is a constant to be determined by the fact that the conic (3) is to pass through the point (2, 2).

$$\therefore\quad 3(4) - 2(2)(2) - 5(4) + 7(2) - 9(2) + (2) + \mu = 0$$

$$\Rightarrow\quad \mu = 18.$$

Putting this value of μ in (3), the required equation of the conic is

Example 36:

Find the lengths and the equations of the axes of the conic $5x^2 - 6xy + 5y^2 + 26x - 22y + 29 = 0$.

Solution:

Let $F(x, y) \equiv 5x^2 - 6xy + 5y^2 + 26x - 22y + 29 = 0$.

The equations giving the coordinates of the centre are

as $$\frac{\partial F}{\partial x} = 10x - 6y + 26 = 0$$

i.e., $5x - 3y + 13 = 0,$ ...(1)

and $\dfrac{\partial F}{\partial y} = -6x + 10y - 22 = 0$ *i.e.,* $3x - 5y + 11 = 0,$ ...(2)

Solving (1) and (2), the coordinates (x_1, y_1) of the centre are (–2, 1). The new constant term are

$$e_1 = gx_1 + fy_1 + c = 13\,(-2) + (-11)\,(1) + 29 = -8.$$

Hence the equation of the conic referred to the centre as origin is second degree terms of the given conic

+ the new constant terms $c_1 = 0$,

$$\Rightarrow \quad 5x^2 - 6xy + 5y^2 - 8 = 0,$$

$$\Rightarrow \quad \tfrac{5}{8}x^2 - \tfrac{6}{8}xy + \tfrac{5}{8}y^2 = 1,$$

which is of the standard form

$$Ax^2 + 2Hxy + By^2 = 1.$$

The quadratic in r^2 giving the squares of the lengths of the semi axes is

$$\left(A - \frac{1}{r^2}\right)\left(B - \frac{1}{r^2}\right) = H^2; \left(\frac{5}{8} - \frac{1}{r^2}\right)\left(\frac{5}{8} - \frac{1}{r^2}\right) = \left(=\frac{3}{8}\right)^2$$

$$\Rightarrow \frac{25}{64} + \frac{1}{r^4} - \frac{5}{4}\cdot\frac{1}{r^2} = \frac{9}{64}, \text{ or } 64 - 80r^2 + 16r^4 = 0$$

$$\Rightarrow r^4 - 5r^2 + 4 = 0, \text{ or } (r^2 - 1)(r^2 - 4) = 0.$$

$$\therefore \quad r_1^2 = 4, r_2^2 = 1.$$

Since both r_1^2 and r_2^2 are +ve, therefore the given conic is an ellipse.

The lengths of the axes of the ellipse are $2r_1$ and $2r_2$ *i.e.,* 4 and 2.

The lengths of the axes of the ellipse are $2r_2$ and $2r_2$ *i.e.,* 4 & 2.

$$\left(A - \frac{1}{r^2}\right)(x - x_1) + H\,(y - y_1) = 0$$

$$\Rightarrow \left(\tfrac{5}{8} - \tfrac{1}{4}\right)(x + 2) + \left(-\tfrac{5}{8}\right)(y - 1) = 0$$

$$\Rightarrow x - y + 3 = 0, \quad \text{or} \quad y = x + 3. \qquad \ldots(3)$$

The equation of the minor axis, which is a line perpendicular to the major axis and passing through the centre (–2, 1) is

$$(y - 1) = -(x + 2)$$

$$\Rightarrow \quad x + y + 1 = 0. \qquad \ldots(4)$$

Thus the equations of the axes of the given conic are given by (3) and (4).

Example 37:

Trace the conic

$$36x^2 + 24xy + 29y^2 - 72x + 126y + 81 = 0.$$

Solution:

The given conic is

$$F(x, y) \equiv 36x^2 + 24xy + 29y^2 - 72x + 126y + 81 = 0.$$

Here $a = 36$, $h = 12$, $b = 29$. Since $h^2 \neq ab$ *i.e.,* the second degree terms are not in a perfect square, therefore the given equation represents a central conic. The coordinates of the centre are given by the equations

$$\partial F/\partial x = 72x + 24y - 72 = 0$$

i.e.,
$$3x + y - 3 = 0$$

and
$$\partial F/\partial y = 24x + 58y + 126 = 0$$

i.e.,
$$12x + 29y + 63 = 0.$$

Solving these equation, *the centre* (x_1, y_1) is the point $(2, -3)$.

$\therefore \quad c_1 = gx_1 + fy_1 + c = (-36) \,.\, (2) + (63)\,(-3) + 81 = -180.$

$\therefore$ the equation of the conic referred to the centre as origin is

$$30x^2 + 24xy + 29y^2 - 180 = 0.$$

In *standard form* $Ax^2 + 2Hy + 8y^2 = 1$ this equation is

$$\tfrac{1}{5}x^2 + \tfrac{2}{15}xy + \tfrac{29}{180}y^2 = 1.$$

$\therefore$
$$A = 1/5,\ B = 29/180,\ H = 1/15.$$

The squares of the lengths of the *semi-axes* are given by

$$\left(A - \frac{1}{r^2}\right)\left(B - \frac{1}{r^2}\right) = H^2,$$

or
$$\left(\frac{1}{5} - \frac{1}{r^2}\right)\left(\frac{29}{180} - \frac{1}{r^2}\right) = \left(\frac{1}{15}\right)^2$$

or
$$\frac{1}{r^2} - \frac{1}{r^2}\left(\frac{1}{5} + \frac{29}{180}\right) + \frac{29}{900} - \frac{1}{225} = 0$$

or
$$\frac{1}{r^4} - \frac{13}{36r^2} + \frac{1}{36} = 0,$$

or
$$r^4 - 13r^2 + 36 = 0$$

or
$$(r^2 - 9)\,(r^2 - 4) = 0.$$

$\therefore$
$$r_1^2 = 9,\ r_2^2 = 4.$$

The conic is *an ellipse* since both r_1^2 and r_2^2 are positive. The lengths of the major and minor axes are $2r_1$ and $2r_2$ *i.e.,* 6 and 4 respectively.

The *equation of the major axis* referred to the centre as origin is

$$\left(A - \frac{1}{r_1^2}\right) x + Hy = 0, \text{ or } \left(\frac{1}{5} - \frac{1}{9}\right) x + \frac{1}{15} y = 0$$

or $\quad 4x + 3y = 0.$

Shifting the origin back, the equation of the *major axis* referred to the old coordinate axes is given by

$$4(x - 2) + 3 \{y - (-3)\} = 3$$

$$\Rightarrow \quad 4x + 3y + 1 = 0.$$

The minor axis is the straight line perpendicular to the major axis and passing through the centre (2, –3). So referred to the old coordinate axes the equation of the *minor axis* is

$$3(x - 2) - 4 (y + 3) = 0$$

$$\Rightarrow \quad 3x - 4y - 18 = 0.$$

[*Remember* that the equation of the line perpendicular to the line $lx + my + n = 0$ and passing through the point (x_1, y_1) is $m (x - x_1) - l (y - y_1) = 0$.]

The points of intersection of the conic with the coordinates axes.

The given conic cuts the x-axis in the points where y = 0 *i.e.,* where $36x^2 - 72x + 81 = 0$,

$$\Rightarrow \quad 4x^2 - 8x + 9 = 0.$$

This gives imaginary values of x because its discriminant

$$b^2 - 4ac = 64 - 4.4.9 = -ve.$$

Hence the given conic does not cut the x-axis.

The given conic cuts the y-axis in the points where x = 0

i.e., $29y^2 + 126y + 81 = 0$

i.e., $y = [-126 \pm \sqrt{\{(126)^2 - 4 \times 29 \times 91\}}]/(58)$

$= -0.8$ and -3.6 (nearly).

Hence the shape of the given conic (which is an ellipse) is as shown in the figure. [See procedure in step 10, 9]. To draw the major axis note that it passes through the centre (2, –3) and it cuts the x-axis at the point

$$(-\tfrac{1}{4}, 0).$$

Example 38:

If t is a parameter, find the locus of the point (x, y) in the following cases:

(*i*) $x = a (t + l/t$ *and* $y = a (t - l/t)$.

(ii) $x = 1 + t + t^2$ and $y = 1 - t + t^2$

(iii) $x = at + bt^2$ and $y = bt + at^2$

(iv) $x = a \tan (t + \alpha)$ and $y = b + \tan (t + \beta)$.

Solution:

The locus of the point (x, y) is obtained by eliminating t between the given relations.

(i) Squarihg, we have

$$x^2 = a^2 \left(t + \frac{1}{t}\right)^2, \; y^2 = a^2 \left(t - \frac{1}{t}\right)^2.$$

But $a^2 (t + 1/t)^2 - a^2 (t - 1/t)^2 = 4a^2$.

$\therefore$ $x^2 - y^2 = 4a^2$ is the required locus.

(ii) We have

$$x + y = 2 (1 + t^2),\; x - y = 2t.$$

$\therefore$ $2 (x + y) = 4 + 4t^2 = 4 + (2t)^2$

$\Rightarrow$ $2 (x + y) = 4 + (x - y)^2$ is the required locus,

(iii) We have

$$ax - by = a^2t + abt^2 - b^2t - abt^2 = (a^2 - b^2)\, t$$

and $$ay - bx = abt + a^2t^2 - abt - b^2t^2 = (a^2 - b^2)\, t^2.$$

$$\therefore \frac{(ax - by)^2}{(ay - bx)} = \frac{(a^2 - b^2)^2\, t^2}{(a^2 - b^2)\, t^2},$$

or $(ax - by)^2 = (a^2 - b^2)(ay - bx)$ is the required locus.

(iv) We any write

$$\tan (\alpha - \beta) = \tan \{(\alpha + t) - (\beta + t)\}$$

$$= \frac{\tan (\alpha + t) - \tan (\beta + t)}{1 + \tan (\alpha + t). \tan (\beta + t)}$$

$$= \frac{x/a - y/b}{1 + (x/a). (y/b)} \quad \text{[from the given relations]}$$

$$= \frac{(bx - ay)}{ab + by}.$$

$\therefore$ $(bx = ay) = (ab + xy) \tan (\alpha - \beta)$ is the required locus.

Example 39:

Frace the curve $8x^2 - 4xy + 5y^2 - 16x - 14y + 17 = 0$.

Find the coordinates of its foci and show that its axes lie along

$$2x - y - 1 = 0 \text{ and } 2x + 4y - 11 = 0.$$

Solution:

The equation of given conic is

$$F(x, y) \equiv 8x^2 - 4xy + 5y^2 - 16x - 14y + 17 = 0.$$

Since the second degree terms are not in a perfect square, therefore the given equation represents a central conic. The co-ordinates (x_1, y_1) of the centre are given by the equations

$$\frac{\partial F}{\partial x} = 16x - 4y - 16 = 0 \text{ and } \frac{\partial F}{\partial y} = -4x + 10y - 14 = 0.$$

Solving these equations, *the centre* is the point $\left(\frac{3}{2}, 2\right)$.

We have $c_1 = gx_1 + fy_1 + c = (-8)\left(\frac{3}{2}\right) + (-7)(2) + 17 = -9.$

$\therefore$ the equation of the conic referred to the centre as origin is

$$8x^2 = 4xy + 5y^2 - 9 = 0.$$

In *standard form* $Ax^2 + 2Hxy + By^2 = 1$ it is

$$\frac{8}{9}x^2 - \frac{4}{9}xy + \frac{5}{9}y^2 = 1.$$

The squares of the *semi-axis* are given by

$$\left(A - \frac{1}{r^2}\right)\left(B - \frac{1}{r^2}\right) = H^2, \text{ or } \left(\frac{8}{9} - \frac{1}{r^2}\right)\left(\frac{5}{9} - \frac{1}{r^2}\right) = \left(-\frac{2}{9}\right)^2$$

$\Rightarrow$ $(8r^2 - 9)(5r^2 - 2) = 4r^4$

$\Rightarrow$ $36r^4 - 117r^2 + 81 = 0,$

or $4r^4 - 13r^2 + 9 = 0$

$\Rightarrow$ $(4r^2 - 9)(r^2 - 1) = 0.$

$\therefore$ $r_1^2 = \frac{9}{4}, r_2^2 = 1.$

Since r_1^2 and r_2^2 are both positive, the conic is an ellipse.

The lengths of the major and minor axes are $2r_1$ and $2r_2$ *i.e.,* 3 and 2 respectively.

The equation of the major axis referred to the centre as origin is

$$\left(A - \frac{1}{r_1^2}\right) x + Hy = 0, \text{ or } \left(\frac{8}{9} - \frac{4}{9}\right) x + \left(-\frac{2}{9}\right) y = 0,$$

$$\Rightarrow \quad 2x - y = 0.$$

Shifting the origin back from the centre $\left(\frac{3}{2}, 2\right)$ to the old origin, the equation of the *major axis* referred to the old coordinate axes is

$$2\left(x - \frac{3}{2}\right) - (y - 2) = 0$$

$$\Rightarrow \quad 2x - y - 1 = 0. \qquad \text{...(1)}$$

The equation of the *minor axis,* which is a line perpendicular to the major axis (1) and passing through the centre $\left(\frac{3}{2}, 2\right)$, is

$$\left(x - \frac{3}{2}\right) + 2(y - 2) = 0, \text{ or } 2x + 4y - 11 = 0 \qquad \text{...(2)}$$

Eccentricity: $e = \sqrt{\left(1 - \frac{r_2^2}{r_1^2}\right)} = \sqrt{\left(1 - \frac{4}{9}\right)} = \frac{\sqrt{5}}{3}.$

Foci : From the equation (1), the slope of the major axis = 2.

$\therefore$ if the major axis makes an angle θ with the x-axis, then $\tan\theta = 2$. Hence $\sin\theta = =2/\sqrt{5}$, $\cos\theta = 1/\sqrt{5}$.

The coordinates of the foci are

$(x_1 + er_1 \cos\theta, y_1 + er_1 \sin\theta)$ and $(x_1 - er_1 \cos\theta, y_1 - er_1 \sin\theta)$

i.e., $$\left(\frac{3}{2} + \frac{\sqrt{5}}{3}.\frac{3}{2}.\frac{1}{\sqrt{5}}, 2 + \frac{\sqrt{5}}{3}.\frac{3}{2}.\frac{2}{\sqrt{5}}\right)$$

and $$\left(\frac{3}{2} - \frac{\sqrt{5}}{3}.\frac{3}{2}.\frac{1}{\sqrt{5}}, 2 - \frac{\sqrt{5}}{3}.\frac{3}{2}.\frac{2}{\sqrt{5}}\right)$$

i.e., (2, 3) and (1, 1).

Points of intersection of the given conic with the coordinate axes

The conic cuts the x-axis in points given by

$$y = 0 \text{ and } 8x^2 - 16x + 17 = 0.$$

The equation $8x^2 - 16x + 17 = 0$ gives imaginary values of x.

[$\because$ $b^2 - 4ac = 256 - 4.8.17 = -$ ive.]

Hence the conic does not cut the x-axis.

The conic cuts the y-axis in points given by

$$x = 0,\ 5y^2 - 14y + 17 = 0,$$

giving imaginary roots.

Hence the curve does not cut the y-axis also.

Hence the shape of the curve is as shown in the figure.

Example 40:

Trace the conic

$$5x^2 + 4xy + 8y^2 - 12x - 12y = 0.$$

Solution:

The given conic is

$$F\ (x, y) \equiv 5x^2 + 4xy + 8y^2 - 12x - 12y = 0.$$

The centre (x_1, y_1) is given by the equations

$$\frac{\partial F}{\partial x} = 10x + 4y - 12 = 0,\quad \frac{\partial F}{\partial y} = 4x + 16y = 12 = 0.$$

Solving, *the centre C* is the point $\left(1, \frac{1}{2}\right)$.

We have $c_1 = gx_1 + fy_1 + c = (-6).1 + (-6).\ \frac{1}{2} + 0 -9.$

The equation of the conic referred to the centre $\left(1, \frac{1}{2}\right)$ as origin is

$$5x^2 + 4xy + 8y^2 - 9 = 0$$

In its *standard form* $Ax^2 + 2Hxy + By^2 = 1$, this equation is

$$\frac{5}{9}x^2 + \frac{4}{9}xy + \frac{8}{9}y^2 = 1.$$

The squares of the semi-axis are given by

$$\left(A - \frac{1}{r^2}\right)\left(B - \frac{1}{r^2}\right) = H^2, \text{ or } \left(\frac{5}{9} - \frac{1}{r^2}\right)\left(\frac{8}{9} - \frac{1}{r^2}\right) = \left(\frac{2}{9}\right)^2$$

$\Rightarrow$ $$36r^2 - 117r^2 + 81 = 0,$$

or $$4r^2 - 13r^2 + 9 = 0.$$

$\Rightarrow$ $$(r^2 - 1)(4r^2 - 9) = 0$$

$\therefore$ $$r_1^2 = \frac{9}{4},\ r_2^2 = 1 \quad \text{i.e., } r_1 = \frac{3}{2},\ r_2 = 1$$

Since r_1^2 and r_2^2 are both + ve, the curve is an ellipse. The lengths of major and minor axes are 3 and 2 respectively.

The equation of the major axis referred to the centre as origin is

$$\left(A - \frac{1}{r_1^2}\right) x + Hy = 0, \text{ or } \left(\frac{5}{9} - \frac{4}{9}\right)x + \frac{2}{9} y = 0,$$

i.e., $x + 2y = 0.$

Shifting the origin back from the centre $\left(1, \frac{1}{2}\right)$ to the old origin, the equation of the *major axis* referred to the old coordinate axes is

$$(x - 1) + 2\left(y - \frac{1}{2}\right) = 0,$$

or $x + 2y = 2 = 0$

Points of intersection of the conic with the coordinate axes. The conic cuts the x-axis in points given by $y = 0$, $5x^2 - 12x = 0$

i.e., in the points (0, 0) and $\left(\frac{12}{5}, 0\right)$.

It meets the y-axis in the points given by $x = 0$,

$$8y^2 = 12y = 0$$

i.e., in the points given by $x = 0$,

$8y^2 - 12y = 0$ *i.e.*, in the points (0, 0) and $(0, \frac{3}{2})$.

Hence the shape of the conic is as shown in the figure.

Example 41:

Trace the curve : $3(3x - 2y + 4)^2 + 2(2x + 3y - 5)^2 = 39.$

Solution:

Proceeding as in Ex. 20 behind, the given equation can be written as

$$3\left\{\frac{3x - 2y + 4}{\sqrt{(9 + 4)}}\right\}^2 + 2\left\{\frac{2x + 3y - 5}{\sqrt{(4 + 9)}}\right\}^2 = 3$$

$$\Rightarrow \quad \frac{\left(\frac{2x + 3y - 5}{\sqrt{13}}\right)}{3/2} + \frac{\left(\frac{3x - 2y + 4}{\sqrt{13}}\right)^2}{1} = 1.$$

Now taking $3x - 2y + 4 = 0$

and $2x + 3y - 5 = 0$ as the new x and y axes respectively, the above equation becomes

$$\frac{X^2}{3/2} + \frac{Y^2}{1} = 1$$

which is the equation of an ellipse whose lengths of major and minor axes are 2.√(3/2) and 2.1 *i.e.,* √6 and 2 respectively.

The equation of the major axis is

Y = 0 *i.e.,* 3x = 2y + 4 = 0 and the equation of the minor axis is

X = 0 *i.e.,* 2x + 2y – 5 = 0

Solving these two equations the centre C is the point $\left(-\frac{2}{13}, \frac{22}{13}\right)$.

The shape is as shown in the figure.

Example 42:

Trace the conic

$$17x^2 - 12xy + 8y^2 + 46x - 28y + 17 = 0.$$

Also find its eccentricity, the equations of its axes, the coordinates of its foci and the equations of its directrics.

Solution:

The given conic is

$$F(x, y) \equiv 17x^2 - 12xy + 8y^2 + 46x - 28y + 17 = 0$$

Since the second degree terms do not form a perfect square, the given equation represents a central conic. The centre (x_1, y_1) is given by the equations

$$\partial F/\partial x = 0$$

i.e., $34x - 12y + 46 = 0$

i.e., $17x - 6y + 23 = 0$

and $\partial F/\partial y = 0$

i.e., $-12y + 16 - 28 = 0$

i.e., $-3x + 4y - 7 = 0.$

Solving these, the centre C is the point (–1, 1).

We have $c_1 = gx_1 + fy_1 + c = 23(-1) + (-14).1 + 17 = -20.$

∴ the equation of the conic referred to the centre as origin

is $$17x^2 - 12xy + 8y^2 - 20 = 0$$

$$\Rightarrow \quad \frac{17}{20}x^2 - \frac{12}{20}xy + \frac{8}{20}y^2 = 1, \text{ reducing to the standard form}$$

$$Ax^2 + 2Hxy + By^2 = 1.$$

The quadratic in r^2 giving the squares of hte semi-axes is

$$\left(A - \frac{1}{r^2}\right)\left(B - \frac{1}{r^2}\right) = H^2, \text{ or } \left(\frac{17}{20} - \frac{1}{r^2}\right)\left(\frac{8}{20} - \frac{1}{r^2}\right) = \left(-\frac{6}{20}\right)^2,$$

$$\Rightarrow \quad (17r^2 - 20)(8r^2 - 20) = 36r^4,$$

$$\text{or} \quad 100r^4 - 500r^2 + 400 = 0.$$

$$\Rightarrow \quad r^4 - 5r^2 + 4 = 0,$$

$$\text{or} \quad (r^4 - 1)(r^2 - 4) = 0.$$

$$\therefore \quad r_1^2 = 4, r_2^2 = 1.$$

Since both the values of r^2 are positive, the given conic is an ellipse.

The length of the major axis $= 2r_1 = 2.2. = 4.$

and the length of the minor axis $= 2r_2 = 2.1 = 2.$

The eccentricity

$$e = \sqrt{\left(1 - \frac{r_2^2}{r_1^2}\right)} = \sqrt{\left(1 - \frac{1}{4}\right)} = \sqrt{\left(\frac{3}{4}\right)} \frac{\sqrt{3}}{2}.$$

The equation of the major axis referred to the old origin is

$$\left(A - \frac{1}{r_1^2}\right)(x - x_1) + H(y - y_1) = 0$$

$$\Rightarrow \quad \left(\frac{17}{20} - \frac{1}{4}\right)\{x - (-1)\} + \left(-\frac{6}{20}\right)(y - 1) = 0$$

$$\Rightarrow \quad \frac{12}{20}(x + 1) - \frac{6}{20}(y - 1) = 0, \text{ or } 2(x + 1) - (y - 1) = 0$$

$$\Rightarrow \quad 2X - Y + 3 = 0. \qquad ...(1)$$

The equation of the *minor axis i.e.,* of the line perpendicular to the major axis (1) and passing through the centre (–1, 1) is

$$\{x - (-1)\} + 2(y - 1) = 0,$$

$$\text{or} \quad x + 2y - 1 = 0. \qquad ...(2)$$

Points of intersection of the given curve with the coordinate axis. The given curve cuts the x-axis where $y = 0$

i.e., where $17x^2 + 46x + 17 = 0$

i.e., where $x = \dfrac{-46 \pm \sqrt{\{(46)^2 - 4 \times 17 \times 17\}}}{16} = \dfrac{-3 \pm \sqrt[4]{15}}{17}$

$= -\ 0.4$ and $-\ 2.3$ nearly.

Again the given curve cuts the y-axis where $x = 0$

i.e., where $8y^2 = 28y + 17 = 0$

i.e., where $x = \dfrac{28 \pm \sqrt{\{(28)^2 - 4 \times 8 \times 17\}}}{16} = \dfrac{7 \pm \sqrt{15}}{4}$

$= 0.8$ and 2.7 nearly.

Hence the shape of the given ellipse is as shown in the figure. To draw the major axis (1), we observe that it passes through the centre C (–1, 1) and cuts the x-axis at the point $\left(-\dfrac{3}{2}, 0\right)$.

Coordinate of the Foci : If θ is the angle which the major 91) makes with the x-axis, we have tan θ = 2 so that

$$\sin\theta = 2/\sqrt{5} \text{ and } \cos\theta = 1/\sqrt{5}.$$

The distance of a focus from the centre C is er_1.

The coordinates of the foci are

$$(x_1 + er_1 \cos\theta,\ y_1 + er_1 \sin\theta)$$

and $(x_1 - er_1 \cos\theta,\ y_1 - er_1 \sin\theta)$

i.e., $$\left(-1 + \frac{\sqrt{3}}{2}.2.\frac{1}{\sqrt{5}},\ 1 + \frac{\sqrt{3}}{2}.2.\frac{2}{\sqrt{5}}\right)$$

and $$\left(-1 - \frac{\sqrt{3}}{2}.2.\frac{1}{\sqrt{5}},\ 1 - \frac{\sqrt{3}}{2}.2.\frac{2}{\sqrt{5}}\right)$$

i.e., $$\left(-1 + \sqrt{\frac{3}{5}},\ 1 + 2\sqrt{\frac{3}{5}}\right)$$

and $$\left(-1 - \sqrt{\frac{3}{5}},\ 1 - 2\sqrt{\frac{3}{5}}\right)$$

Equations of the Directrices : The distance of a directrix from the centre

$$C = \frac{r_1}{e} = \frac{2}{\sqrt{3/2}} = \frac{4}{\sqrt{3}}$$

The directrices are the straight lines perpendicular to the major axis and at a distance r_1/e from the centre on either side of it. So the equations of the directrices are

$$(x_1 - x_1)\cos\theta + (y - y_1)\sin\theta = \pm r_1/e$$

$$\Rightarrow \quad (x+1)\cdot\frac{1}{\sqrt{5}} + (y-1)\cdot\frac{2}{\sqrt{5}} = \pm\frac{4}{\sqrt{3}}$$

$$\Rightarrow \quad (x+1) + 2(y-1) = \pm 4\sqrt{\frac{5}{3}}$$

$$\Rightarrow \quad x + 2y = 1 \pm 4\sqrt{\frac{5}{3}}.$$

Alternative method for finding the equations of the directrices.

The equation of any line perpendicular to the major axis (1) is

$$x + 2y = \lambda \qquad ...(3)$$

For (3) to be the equation of a directrix the length of the perpendicular from the centre (1, 1) to the line (3) must be equal to $\pm r_1/e$

i.e., $\dfrac{-1+2-\lambda}{\sqrt{5}} = \pm\dfrac{4}{\sqrt{3}}$

i.e., $1-\lambda = 1 \pm 4\sqrt{\dfrac{5}{3}}$

i.e., $\lambda = 1 \pm 4\sqrt{\dfrac{5}{3}}$

Putting this value of λ in (3), the equation of the directrices are

$$x + 2y = 1 \pm 4\sqrt{\frac{5}{3}}$$

Example 43:

Trace $7x^2 + 52xy - 32y^2 - 170x + 140y = 0$ and find the equation of its asymptotes.

Solution:

The given conic is

$$F(x, y) \circ 7x^2 + 52xy - 32y^2 - 170x + 140 = 0.$$

The centre (x_1, y_1) is given by the equations

$$\frac{\partial F}{\partial x} = 14x + 52y - 170 = 0, \quad \frac{\partial F}{\partial y} = 52x - 64y + 140 = 0$$

Solving these, *the centre* C is the point (1, 3).

We have $c_1 = gx_1 + f_1 + c = (-85).(1) + (70).(3) + 0 = 125.$

The equation of hte conic referred to the centre (1, 3) as origin is

$$7x^2 + 52xy - 32y^2 + 125 = 0$$

$$\Rightarrow \qquad -\frac{7}{125}x^2 - \frac{52}{125}xy + \frac{33}{125}y^2 = 1,$$

reducing to the standard form $Ax^2 + 2Hxy + By^2 = 1$.

The squares of the *semi-axes* are given by

$$\left[A - \frac{1}{r^2}\right]\left[B - \frac{1}{r^2}\right] = H^2, \text{ or } \left[-\frac{7}{125} - \frac{1}{r^2}\right]\left[\frac{32}{125} - \frac{1}{r^2}\right] = \left[-\frac{26}{125}\right]^2$$

$$\Rightarrow \qquad 36r^2 + 125r^2 - 625 = 0,$$

or $\qquad (9r^2 - 25)(4r^2 + 25) = 0.$

$$\therefore \qquad r_1^2 = 25/9,\ r_2^2 = -25/4.$$

Since r_1^2 and r_2^2 are of opposite signs, the curve is a hyperbola. The lengths of the transverse and conjugate axes are 10/3 and 5 respectively.

4

Intersection of a Curve and a Line

INTRODUCTION

Let the equation of the curve be given by

$$ax^2 + 2hxy + by^2 + 2gx + 2fy + c = 0 \quad \text{...(i)}$$

and the equ. of line is given as

$$lx + my = 1. \quad \text{...(ii)}$$

Let the curve and the line intersect at the points A and B. Now the equation

$$ax^2 + 2hxy + by^2 + 2gx + 2hy + c\,(lx + my)^2 = 0 \quad \text{...(iii)}$$

represents same curve passing through A and B. But the equation being homogenous in x and y of second degree represents pair of straight line passing through the origin and the points of inter section of a curve and a straight line. Thus the equation of the pairs of lines joining the origin and the points of intersection of a line and a curve is obtained by making the curve homogenous with the help of the line.

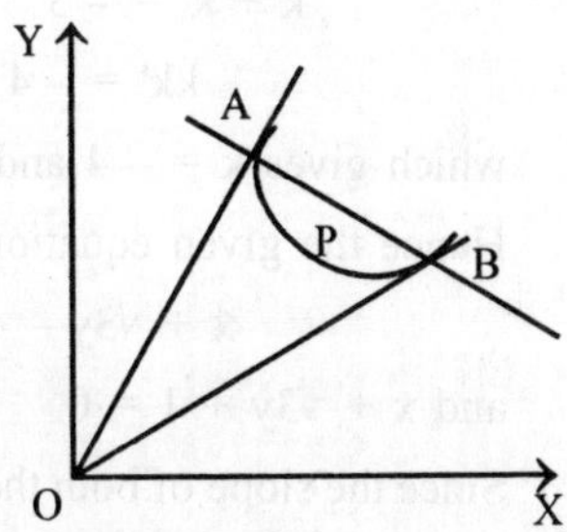

Example 1:

Show that the equation $x^2 + 2\sqrt{3}\,xy + 3y^2 - 3x - \sqrt{3}\,y - 4 = 0$ represents two parallel straight lines and find the distance between them.

Solution:

Comparing the given equation with general equation of second degree $ax^2 + 2hxy + by^2 + 2gx + 2fy + c = 0$, we have

$$a = 1,\ h = \sqrt{3},\ b = 3,\ g = -\frac{3}{2}\ f = -\frac{3}{2}\sqrt{3} \text{ and } c = -4.$$

Substituting these values in the required condition for pair of straight lines, i.e., in $abc + 2fgh - af^2 - bg^2 - ch^2$, we get

$$-12 + 2\left(-\frac{3}{2}\right)\sqrt{3}\,.\left(-\frac{3}{2}\right)\sqrt{3} - \frac{27}{4} - 3\left(\frac{9}{4}\right) + 4.3$$

$$= -12 + 2\,.\,\frac{27}{4} - \frac{27}{4} - \frac{27}{4} + 12$$

$$= 0.$$

This shows that the equation represents a pair of straight lines the given equation can be written as

$$(x + \sqrt{3}\ y)^2 - 3x - 3\sqrt{3}\ y - 4 = 0.$$

Let it represents the lines

$$x + \sqrt{3}\ y + k = 0$$

and $x + \sqrt{3}\ y + k' = 0.$

This gives

$$(x + \sqrt{3}y)^2 - 3x - 3\sqrt{3}y - 4 \equiv (x + \sqrt{3}y + k)(x + \sqrt{3}y + k')$$

Comparing and equating coefficients of the like terms, we get

$$k + k' = -3$$

$$kk' = -4$$

which gives $k = -4$ and $k' = 1$.

Hence the given equation represents the straight lines

$$x + \sqrt{3}y - 4 = 0$$

and $x + \sqrt{3}y + 1 = 0$.

Since the slope of both the lines is same, there fore, the lines are parallel. The distance between these parallel lines is the length of the prependicular distance from a point of one line to the other line. Clearly, $(-1, 0)$ lies on the line $x + \sqrt{3}y + 1 = 0$. Therefore, the perpendicular distance from $(-1, 0)$ to $x + \sqrt{3}y - 4 = 0$ is $\dfrac{-1-4}{\sqrt{(1+3)}} = \dfrac{-5}{2} = \dfrac{5}{2}$ (negative is dropped for distance)

Example 2:

Find the equation to the pair of straight lines joining the origin to the intersection of the straight line $y = mx + c$ and the curve $x^2 + y^2 = a^2$. Prove *that they are at right angles if*

$$2c^2 = a^2(1 + m^2)$$

Solution:

The equation of the curve is $x^2 + y^2 = a^2$ ...(i)

The equation of the straight line is $y = mx + c$ ...(ii)

Making (i) homogeneous with the help of (ii), we have

$$x^2 + y^2 = a^2\left(\frac{y - mx}{c}\right)^2$$

$$\Rightarrow \; x(c^2 - a^2m^2) = 2a^2mxy + y^2(c^2 - a^2) = 0. \qquad ...(iii)$$

The equation (iii) being homogeneous will give two straight lines.

These lines will be at right angles if

coefficient of x^2 + coefficient of $y^2 = 0$.

$\Rightarrow \; c^2 - a^2m^2 + c^2 - a^2 = 0$

$\Rightarrow \; 2c^2 = a^2(1 + m^2)$

Hence proved.

LENGTH OF PERPENDICULAR

To find the length of the perpendicular let fall from a given point upon a given straight line.

Case I. *When the equation of the given straight line is*

$$x \cos \alpha + y \sin \alpha - p = 0$$

Let AB be a straight line and $P(x_1, y_1)$ be any point. From P draw a line CD parallel to the given line and let P' (OF) be the perpendicular distance from O on CD. The equation of CD is given by

$$x \cos \alpha + y \sin \alpha = p'$$

This line passes through the point $P(x_1, y_1)$.

$\therefore \; x_1 \cos \alpha + y_1 \sin \alpha = p'$

FE = perpendicular distance from P on the line AB

$= FO - EO$

$= p' - p$

$= x_1 \cos \alpha + y_1 \sin \alpha = p.$

Hence, the length of the perpendicular from a point (x_1, y_1) on the line $x \cos \alpha + y \sin \alpha = p$ is given as

$$x_1 \cos \alpha + y_1 \sin \alpha - p$$

Case II. *When the equation of the given straight line is given by*

$$ax + by + c = 0.$$

The equation of the given straight line is

$$ax + by + c = 0$$

Reducing it to perpendicular form by dividing throughout by $\sqrt{a^2 + b^2}$ we have

$$\frac{ax}{\sqrt{a^2+b^2}} + \frac{by}{\sqrt{a^2+b^2}} + \frac{c}{\sqrt{a^2+b^2}} = 0$$

Now applying the result obtained in case I, we find the length of the perpendicular from the point (x_1, y_1) to the given line $ax + by + c$ is

$$\frac{ax_1 + by_1 + c}{\sqrt{a^2+b^2}}$$

Example:

If p and p' are the perpendicular from the origin on the straight lines whose equations are

$$x \sec \theta - y \operatorname{cosec} \theta = a$$

and $\quad x \cos \theta + y \sin \theta = a \cos q2\theta,$

prove that $\quad 4p^2 + p'^2 = a^2.$

Solution:

Length of perpendicular from (0, 0) to the line

$x \sec \theta - y \operatorname{cosec} \theta - a = 0$ is givun as

$$p = \frac{a}{\sqrt{\sec^2 \theta + \operatorname{cosec}^2 \theta}}$$

$$\Rightarrow \quad p^2 = \frac{a^2}{\dfrac{1}{\cos^2 \theta} + \dfrac{1}{\sin^2 \theta}} = \frac{a^2 \sin^2 \theta \cos^2 \theta}{\sin^2 \theta + \cos^2 \theta}$$

$$\therefore \quad p^2 = a^2 \sin^2 \theta \cos^2 \theta \qquad (\because \sin^2 \theta + \cos^2 \theta = 1)$$

Also length of perpendicular from (0, 0) to the line is

$x \cos\theta + y \sin\theta - a \cos 2\theta = 0$ is

$$p' = \frac{-a\cos 2\theta}{\sqrt{\cos^2\theta + \sin^2\theta}} = -a\cos 2\theta$$

$\therefore \quad p'^2 = a^2 \cos^2 2\theta$

$\therefore \quad 4p^2 + p'^2 = 4a^2 \sin^2\theta \cos^2\theta + a^2 \cos^2. 2\theta$

$= a^2(\sin^2 2\theta + \cos^2 2\theta) = a^2$

POSITION OF POINT WITH RESPECT TO A LINE

To show that two points (x_1, y_1) and (x_2, y_2) are on the same side of the line $ax + by + c = 0$ if the expression $ax_1 + by_1 + c$ and $ax_2 + by_2 + c$ have the same sign and on the opposite sides if the expressions are of opposite sign.

Proof:

Let AB be the straight line $ax + by + c = 0$ and $P(x_1, y_1)$ and $Q(x_2, y_2)$ be two points. Let PS and QS' be the perpendiculars on x-axis meeting the line at R and R'. Since R and R' have the same abscissas as P and Q respectively the co-ordinates of and R' can be taken as (x_1, y_1) and $R(x_2, y_2')$ respectively

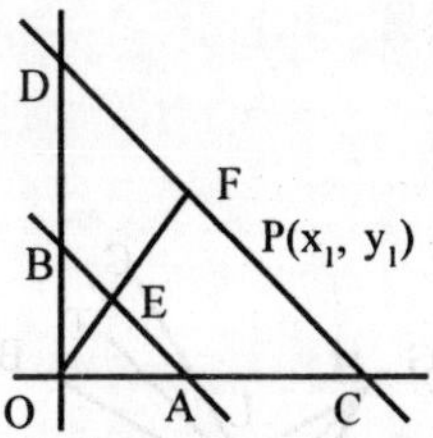

Case I. P, Q lie on same side

(i) above the line

$\Rightarrow \quad y_1 > y_1'$ and $y_2 > y_2'$

Also $ax_1 + by_1' + c = 0$ and $ax_2 + by_2' + c = 0$ (Since R and R' lie on AB)

$\therefore \quad ax_1 + by_1 + c > ax_1 + by_1' + c = 0$

and $ax_2 + by_2 + c > ax_2 + by_2' + c = 0$

$\therefore$ Both the expressions have the same sign.

(ii) below the line.

In this case $y_1 < y_1'$ and $y_2 < y_2'$ both the expressions $ax_1 + by_1 + c$ and $ax_2 + by_2 + c$ will have negative sign. Thus in both the cases the expressions have the same sign.

Case II. P, Q lie on opposite side

$\Rightarrow \quad y_1 < y_1', y_2 < y_2'$

or $\quad y_1 < y_1', y_2 < y_2'$

$\Rightarrow \quad ax_1 + by_1 + c > ax_1 + by_1' + c > 0$

and $ax_2 + by_2 + c < ax_2 + by_2' + c < 0$

or $\quad ax_1 + by_1 + c < ax_1 + by_1' + c < 0$

and $ax_2 + by_2 + c > ax_2 + by_2' + c > 0$

Thus in both the possibilities the expressions have opposite signs.

Corollary 1. The point (x_1, y_1) and $(0, 0)$ lie on the same side of the line $ax + by + c = 0$ if $ax_1 + by_1 + c > 0$.

BISECTORS OF ANGLES

To find the equation of the straight lines which bisect the angles between two given straight lines.

Let AB and CD be given lines intersecting at E. Let their equations be given by

$$a_1x + b_1y + c_1 = 0 \qquad ...(i)$$

$$a_2x + b_2y + c_2 = 0 \qquad ...(ii)$$

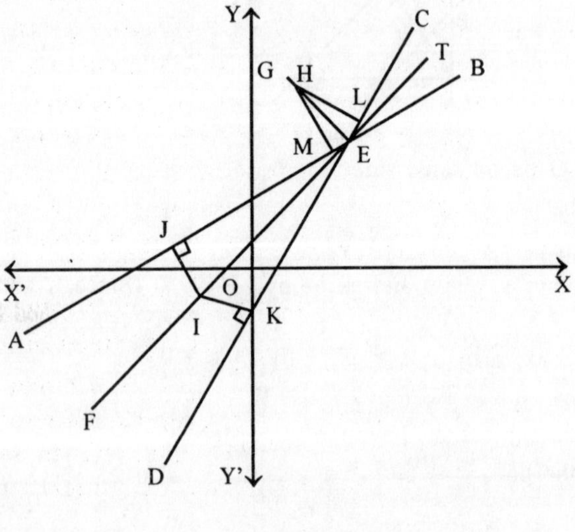

Let EF and EG be the bisectors of angles ∠AED and ∠CEA. Take any two points I and H on EF and EG respectively. From I draw perpendiculars IJ and IK on AB and CD respectively. From H draw HL and HM perpendiculars on CD and AB respectively.

Points I and H are on the same side of CD

⇒ IK and HL have the same sign.

Points I and H are on the opposite sides of AB

⇒ IJ and HM have opposite signs.

⇒ If IJ and IK have same sign then HL and HM have opposite signs and vice versa.

By simple properties of a triangle

IJ = IK

HL = HM.

If (x, y) denote the co-ordinate of I or H then using the perpendicular distance formula we have the equation of EF as

$$\frac{a_1x+b_1y+c_1}{\sqrt{a_1^2+b_1^2}} = \frac{a_2x+b_2y+c_2}{\sqrt{a_2^2+b_2^2}} \quad ...(i)$$

and the equation of EG

$$\frac{a_1x+b_1y+c_1}{\sqrt{a_1^2+b_1^2}} = -\frac{a_2x+b_2y+c_2}{\sqrt{a_2^2+b_2^2}} \quad ...(ii)$$

This gives the required equation of the bisectors as

$$\frac{a_1x+b_1y+c_1}{\sqrt{a_1^2+b_1^2}} = \pm\frac{a_2x+b_2y+c_2}{\sqrt{a_2^2+b_2^2}}$$

DISTINCTION BETWEEN THE TWO BISECTORS WITH THE HELP OF THE ORIGIN

Let the equations of the two lines be given by

$$u \equiv a_1x + b_1y + c_1 = 0$$

and $v \equiv a_2x + b_2y + c_2 = 0$

where c_1 and c_2 are both positive.

Then the origin lies on the positive side of each of the lines.

Let $P_1(x_1, y_1)$ be any point on the bisector which contains the origin. Since P_1 and O lie on the same side of each line the perpendicular from P_1 on u

= 0, v = 0 must be positive. Also since c_1 and c_2 are positive, there fore $a_1x_1 + b_1y_1 + c_1$ and $a_2x_2 + b_2y_2 + c_2$ are also positive. Hence the bisector which contains the origin is

$$\frac{a_1x+b_1y+c_1}{\sqrt{a_1^{\,2}+b_1^{\,2}}}=\frac{a_2x+b_2y+c_2}{\sqrt{a_2^{\,2}+b_2^{\,2}}}$$

LINES THROUGH THE POINT OF INTERSECTION OF TWO STRAIGHT LINES

Show that the equation of any straight line passing through the intersection of straight line $u = 0$ and $v = 0$ is $u + \lambda v = 0$, where λ is any arbitrary constant.

Proof:

Let the equation of the straight lines be given by

$$u \equiv a_1x + b_1y + c_1 = 0 \qquad \text{...(i)}$$

$$v \equiv a_2x + b_2y + c_2 = 0 \qquad \text{...(ii)}$$

Let (h, k) be the point of intersection of lines (i) and (ii). It will be shown that (h, k) lies on the line

$$a_1x + b_1y + c_1 + \lambda\ (a_1x + b_1y + c_1) = 0. \qquad \text{...(iii)}$$

Since (h, k) is the point of intersection of lines (i) and (ii), we have

$$a_1h + b_1k + c_1 = 0. \qquad \text{...(iv)}$$

$$a_2h + b_2k + c_2 = 0. \qquad \text{...(v)}$$

Now the expression

$$(a_1h + b_1k + c_1) + \lambda\ (a_2h + b_2k + c_2) = 0$$

for all values of λ.

This shows that (h, k) lies on line (iii)

Thus whatsoever be the values of λ, equation (iii) represents a straight line passing through the intersection of (i) and (ii).

CONDITION FOR THE CONCURRENCY OF THREE STRAIGHT LINES

Let the three straight lines be

$$a_1x + b_1y + c_1 = 0$$

$$a_2x + b_2y + c_2 = 0$$

$a_3x + b_3y + c_3 = 0$

and let these lines meet at the point (h, k). Then we have

$a_1h + b_1k + c_1 = 0$

$a_3h + b_2k + c_2 = 0$

$a_3h + b_3k + c_3 = 0.$

Eliminating h and k between these three equations, we get

$$\begin{vmatrix} a_1 & b_1 & c_1 \\ a_2 & b_2 & c_2 \\ a_3 & b_3 & c_3 \end{vmatrix} = 0,$$

which is the required condition for the concurrency of three straight lines.

Another test for concurrency of three lines

Let the three lines be

$A_1x + B_1y + C_1 = 0$ ...(i)

$A_2x + B_2y + C_2 = 0$ ...(ii)

$A_3x + B_3y + C_3 = 0$ ...(iii)

Definition: *If there constants p, q, r, can be found and such that*

$p(A_1x + B_1y + C_1) + q(A_2x + B_2y + C_2) + r(A_3x + B_3y + C_3) \equiv 0$ ...(iv)

Identically, (i.e., expression vanishes irrespective of the values of x and y) then the three lines must meet in a point.

Proof:

Let (α, β) be the point of intersection of (i) and (ii), then

$A_1\alpha + B_1\beta + C_1 = 0$ ...(v)

$A_2\alpha + B_2\beta + C_2 = 0$ ...(vi)

Now (iv) is an identity in x and y and hence true for all values of x and y. Putting x = α and y = β in (iv), we have

$$p(A_1\alpha + B_1\beta + C_1) + q(A_2\alpha + B_2\beta + C_2) + r(A_3\alpha + B_3\beta + C_3) \equiv 0$$

$$A_3\alpha + B_3\beta + C_3 = -\frac{p}{r}(A_1\alpha + B_1\beta + C_1) - \frac{q}{r}(A_2\alpha + B_2\beta + C_2)$$

$$= -\frac{p}{r}.0 - \frac{q}{r}.\ 0 = 0 \qquad \text{[By (v) and (vi)]}$$

which show that (α, β) is also a point on the line (iii).

Hence, the three lines are concurrent.

Example 1:

Show that the lines $lx + my + n = 0$, $mx + ny + l = 0$ and $nx + ly + m = 0$ are concurrent if $l + m + n = 0$.

Solution:

The equation of the lines are

$$lx + my + n = 0$$

$$mx + ny + l = 0$$

$$nx + ly + m = 0.$$

The lines are concurrent if

$$\begin{vmatrix} l & m & n \\ m & n & l \\ n & l & m \end{vmatrix} = 0$$

Operating $C_1 + (C_2 + C_3)$, we have

$$\begin{vmatrix} l+m+n & m & n \\ l+m+n & n & l \\ l+m+n & l & m \end{vmatrix} = 0$$

or $$(l + m + n)\begin{vmatrix} 1 & m & n \\ 1 & n & l \\ 1 & l & m \end{vmatrix} = 0$$

which holds as $l + m + n = 0$.

Example 2:

Prove analytically that altitude of a triangle meet at a point.

Solution:

Let $A(x_1, y_1)$ $B(x_2, y_2)$ and $C(x_3, y_3)$ be the vertices of a triangle. Draw AD is perpendicular from A on BC.

Now slope of $BC = \dfrac{y_3 - y_2}{x_3 - x_2}$

$\therefore$ Slope of $AD = \dfrac{(x_3 - x_2)}{y_3 - y^2}$

$\therefore$ equation of AD is $y - y_1 = -\dfrac{x_3 - x_2}{y_3 - y_2}(x - x_1)$

which on simplification becomes

$x(x_3 - x_2) + y(y_3 - y_2) - y_1(y_3 - y_2) - x_1(x_3 - x_2) = 0$ ...(i)

similarly the equation of altitudes BE and CF are

$x(x_1 - x_3) + y(y_1 - y_3) - y_2(y_1 - y_3) - x_2(x_1 - x_3) = 0$ ...(ii)

and $x(x_2 - x_1) + y(y_2 - y_1) - y_3(y_2 - y_1)\ x_3(x_2 - x_1) = 0$ ...(iii)

These lines will be concurrent if

$$\begin{vmatrix} x_3 - x_2 & y_3 - y_2 & y_1(Y_3 - y_2) + x_1(x_3 - x_2) \\ x_1 - x_3 & y_1 - y_3 & y_2(Y_1 - y_3) + x_2(x_1 - x_3) \\ x_2 - x_1 & y_2 - y_1 & y_3(Y_2 - y_1) + x_3(x_2 - x_3) \end{vmatrix} = 0$$

Operating $R_1 + R_2 + R_3$ we get

$$\begin{vmatrix} 0 & 0 & 0 \\ x_1 - x_3 & y_1 - y_3 & y_2(y_1 - y_3) + x_2(x_1 - x_3) \\ x_2 - x_1 & y_2 - y_1 & y_3(y_2 - y_1) + x_3(x_2 - x_3) \end{vmatrix} = 0.$$

Hence the lines are concurrent

Example 3:

A triangle is formed by the lines $y + x - 6 = 0$, $3y - x + 2 = 0$, $5x + 3y + 2 = 0$. Find the coordinates of its orthocentre.

Solution:

Let ABC be the given triangle the equations of whose sides BC, CA and AB are

$x + y - 6 = 2$ ...(i)

$-x + 3y + 2 = 0$ ...(ii)

$5x - 3y + 2 = 0.$ ...(iii)

The coordinates of the point of intersection of the lines AB and AC are given by

$$\frac{x}{-6-6} = \frac{y}{-10-2} = \frac{1}{-3+15}$$

$\Rightarrow$ x = – 1, y = – 1.

$\therefore$ the coordinates of A are (–1, – 1).

The coordinates of the point of intersections of the lines BC and AB are given by

$$\frac{x}{2-18} = \frac{y}{-30-2} = \frac{1}{-3-5}.$$

$\therefore$ x = 2. y = 4.

$\therefore$ the coordinates of B are (2, 4).

The slope of BC is –1 and the slope of altitude AL is 1.

$\therefore$ the equation of the line AL is

$$y + 1 = 1(x + 1)$$

or $x - y = 0$ ...(iv)

Slope of altitude BM is –3.

$\therefore$ the equation of the line BM is

$$y - 4 = -3(x - 2)$$

$\Rightarrow$ $3x + y - 10 = 0.$ ...(v)

Solving (iv) and (v), we get $x = \frac{5}{2}$ and $y = \frac{5}{2}$.

$\therefore$ the coordinates of the orthocentre H are $\left(\frac{5}{2}, \frac{5}{2}\right)$.

JOINT EQUATION FOR MORE THAN ONE STRAIGHT LINE

Consider the equation

$$(a_1x + b_1y + c_1)(a_2x + b_2y + c_2) = 0. \quad ...(i)$$

Any point (h, k) which satisfies $a_1x + b_1y + c_1 = 0$ will also satisfy equation (i) similarly, every point on the line $a_2x + b_2y + c_2 = 0$ also satisfies equation (i). Thus equation (i) is satisfied by any point on either of the given lines, and conversely, on point which does not satisfy any of the lines cannot satisfy equation (i). Thus (i) represents the joint equation of the straight lines

$$a_1x + b_1y = c_1 = 0$$

$$a_2x + b_2y = c_2 = 0$$

Homogenous Equation

An equation of the type

$$a_0y^n + a_1y^{n-1}x + a_2y^{n-2}x^2 + ...a_nx^n = 0 \quad ...(i)$$

is called an homogenous equation of degree n, since the sum of powers of indices of x and y in every term is the same and equal to n.

Dividing equation (i) by x^n, we have

$$a_0\left(\frac{y}{x}\right)^n + a_1\left(\frac{y}{x}\right)^{n-1} + a_2\left(\frac{y}{x}\right)^{n-2} + ... + a_n = 0.$$

This equation being nth degree in y/x will have n roots. If m_1, m_2, m_3, ..., m_n are the roots of the above equation, then equation (i) can be expressed as

$$a_0\left(\frac{y}{x}-m_1\right)\left(\frac{y}{x}-m_2\right)\left(\frac{y}{x}-m_3\right)......\left(\frac{y}{x}-m_n\right)=0$$

which implies that

$$y - m_1x = 0,\ y - m_2x = 0,\ m_3x = 0,\ ......,\ m_nx = 0$$

All these equation represent straight lines passing through the origin.

Thus a homogeneous equation of degree n represents n straight lines passing through the origin.

ANGLES BETWEEN THE LINES

To find the angle between the two straight lines represented by the equation

$$ax^2 + 2hxy + by^2 = 0$$

Proof:

Let the lines represented by this equation be

$$y - mx = 0$$

and $y - m'x = 0$

The joint equation of these two straight lines is given by

$$(y - mx)(y - m'x) = 0.$$

$$\therefore \quad y^2 + \frac{2h}{b}xy + \frac{a}{b}x^2 \equiv (y - mx)(y - m'x)$$

Comparing the co-efficients of like terms, we have

$$m + m' = -2h/b$$

$$mm' = b/a$$

Let θ represent the angles between the straight lines, the we have

$$\tan\theta = \frac{m - m'}{1 + m_1 m_2}$$

$$\frac{\sqrt{(m+m')^2 - 4mm'}}{1 + m_1 m_2}$$

$$\frac{\sqrt{\left(\frac{-2h}{b}\right)^2 - 4\frac{b}{a}}}{1 + b/a}$$

$$\frac{\sqrt{4h^2 - 4ab}}{b+a}$$

$\therefore$ The angle between the lines $ax^2 + 2hxy + by^2 = 0$ is

$$\theta = \tan^{-1} \frac{2\sqrt{h^2 - ab}}{b+a}$$

Corollary 1. *Condition for perpendicularity.*

The lines are perpendicular if $\theta = 90°$. The gives $a + b = 0$.

Corollary 2. *Condition for coincidence or parallel lines.*

The lines are coincidence if $\theta = 0$, i.e. if $h^2 = ab$

BISECTOR FORMULA

To find the equation of bisectors of the angle between the two straight lines represented by the equation

$$ax^2 + 2hxy + by^2 = 0.$$

Proof:

Let the equations of lines represented by $ax^2 + 2hxy + by = 0$ be

$y = mx$ and $y = m'x$. Then we have

$$\left.\begin{aligned} m + m' &= \frac{-2h}{b} \\ mm' &= a/b \end{aligned}\right\} \quad \text{...(i)}$$

The equation of the bisectors of the angle between the lines are $y = mx$ and $y = m'x$ is given by

$$\frac{y - mx}{\sqrt{1+m^2}} = \pm \frac{y - m'x}{\sqrt{1+m'^2}}.$$

This gives the combined equation of the bisectors as

$$\left(\frac{y-mx}{\sqrt{1+m^2}}-\frac{y-m'x}{\sqrt{1+m'^2}}\right)\left(\frac{y-mx}{\sqrt{1+m^2}}+\frac{y-m'x}{\sqrt{1+m'^2}}\right)=0$$

$$\Rightarrow \quad \frac{(y-mx)^2}{1+m^2}-\frac{(y-m'x)^2}{1+m'^2}=0.$$

$$\Rightarrow \quad y^2(m'^2 - m^2) + 2xy(m' - m)(1 - mm') + x^2(m^2 - m'^2) = 0$$

$$\Rightarrow \quad y^2(m + m') + 2xy(1 - mm') + x^2(m + m') = 0 \text{ (as } m_1 \neq m'_2)$$

$$\Rightarrow \quad y^2 + 2xy\left(\frac{1-mm'}{m+m'}\right)-x^2=0.$$

$$\Rightarrow \quad y^2 - x^2 = -2xy\left\{\frac{1-a/b}{-2h/b}\right\} \text{ [substituting } m + m' \text{ and } mm' \text{ from (i)]}$$

we have $y^2 - x^2 = \dfrac{b-a}{h}xy$

$$\Rightarrow \quad \frac{x^2-y^2}{a-b}=\frac{xy}{h}$$

Hence the equation of the bisectors of the angle between the lines

$$ax^2 + 2hxy + by^2 = 0 \text{ is}$$

$$\frac{x^2-y^2}{a-b}=\frac{xy}{h} \text{ given by.}$$

CONDITION THAT A GENERAL EQUATION OF SECOND DEGREE REPRESENTS PAIR OF STRAIGHT LINES

The equation $ax^2 + 2hxy + by^2 + 2gx + 2fy + c = 0$ is called the general equation of second degree as it contains all the terms not only of second degree but also terms of lower degree. We now obtain the condition that general equation of second degree represents two straight lines.

Let $ax^2 + 2hxy + by^2 + 2gx + 2fy + c = 0$ represents the two straight lines.

$$lx + my + n = 0 \quad \text{and} \quad l'x + m'y + n' = 0.$$

Then we have

$$ax^2 + 2hxy + by^2 + 2gx + 2fy + c \equiv (lx + my + n)(l'x + m'y + n')$$

Comparing the co-efficients of the like terms. We have

$$\frac{ll'}{a}=\frac{lm'+l'm}{2h}=\frac{mm'}{b}=\frac{ln'+l'n}{2g}=\frac{mn'+m'n}{2f}=\frac{nn'}{c}=\lambda(\text{say})$$

$\Rightarrow\ ll' = a\lambda$ (i); $lm' + l'm = 2h\lambda$ ...(iv)

$mm' = b\lambda$ (ii); $ln' + l'n = 2g\lambda$ ...(v)

$nn' = c\lambda$ (iii); $mn' + m'n = 2f\lambda$ ...(vi)

Multiplying the relations (iv), (v) and (vi), we have

$$8fgh\,\lambda^3 = (lm' + l'm)\,(ln' + l'n)\,(mn' + m'n)$$

$$= 2\,ll'\,mm'\,nn' + ll'\,(m'^2n^2 + m^2n'^2) + mm'\,(n^2l'^2 + n'^2l^2) + nn'\,(l^2m'^2 + l'^2m^2)$$

$$= 2abc\,\lambda^3 + a\lambda\,(4f^2\lambda^2 - 2bc\,\lambda^2) + b\,\lambda\,(4g^2\lambda^2 - 2ac\,\lambda^2) + c\,\lambda\,(4h^2\lambda^2 - 2ab\,\lambda^2)$$

$$\Rightarrow\ 8fgh = -4abc + 4af^2 + 4bg^2 + 4ch^2$$

$$\Rightarrow\ abc + 2fgh - af^2 - bg^2 - ch^2 = 0.$$

which is the required condition. That a general equation of second degree represent pair of straight this condition also be continue as

$$\begin{vmatrix} a & g & h \\ g & b & f \\ h & f & c \end{vmatrix} = 0.$$

ANGLE BETWEEN THE LINES REPRESENTED BY THE EQUATION

$$ax^2 + 2hxy + by^2 + 2gx + 2fy + c = 0$$

Proof:

Let $l_1x + m_1y + n_1 = 0$ ...(i)

$l_2x + m_2y + n_2 = 0$ ...(ii)

be the lines represented by the given equation. we know that

$$l_1l_2 = a\lambda,\ m_1m_2 = b\lambda,\ n_1n_2 = c\lambda,\ l_1m_2 + l_2m_1 = 2h\lambda,$$

$$m_1m_2 + m_2n_1 = 2f\lambda \text{ and } l_1n_2 + l_2n_1 = 2g\lambda,$$

the slope of lines (i) and (ii) are $-l_1/m_1$ and $-l_2/m_2$.

Let θ be the angle between them.

$$\therefore\ \tan\theta = \frac{l_1/m_1 - l_2/m_2}{l_1l_2\ m_1m_2} = \frac{l_1m_2 - l_2m_1}{l_1l_2 + m_1m_2}$$

$$= \frac{\sqrt{(l_1 m_2 + l_2 m_1)^2 - 4 l_1 l_2 m_1 m_2}}{l_1 l_2 + m_1 m_2}$$

$$= \frac{2\lambda\sqrt{h^2 - ab}}{\lambda(a+b)}$$

$$= \frac{2\sqrt{h^2 - ab}}{a+b}$$

Which is the required formula for calculating the angle between the two lines represented by the general equation of second degree. Condition for perpendicular is a + b = 0 condition for parallelogram is $h^2 = ab$

TWO OR MORE STRAIGHT LINES

Angle Between Two Straight Lines

To find the angle between given straight lines

$$y = m_1x + c_1$$

and $y = m_2x + c_2$.

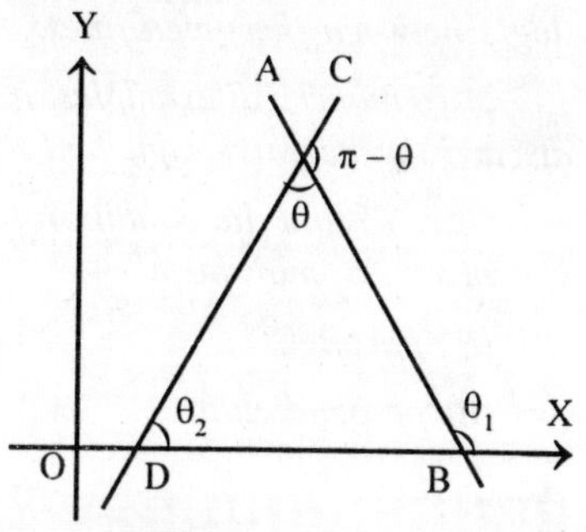

Proof :

Let AB and CD be two given straight lines whose equations are given by

$$y = m_1x + c_1 \text{ and } y = m_2x + c_2$$

respectively. Let P be the point of intersection of AB and CD, so that if θ_1 and θ_2 are the angles they make with the positive direction of x-axis then

$$\tan\theta_1 = m_1 \text{ and } \tan\theta_2 = m_2 \qquad \text{...(i)}$$

Let θ be the required angle between then we have two lines then

$$\theta_1 = \theta + \theta_2$$

or $$\theta = \theta_1 - \theta_2.$$

$\therefore$ $$\tan\theta = \tan(\theta_1 - \theta_2)$$

$$= \frac{\tan\theta_1 - \tan\theta_2}{1 + \tan\theta_1 \tan\theta_2}$$

$$= \frac{m_1 - m_2}{1 + m_1 m_2} \qquad \text{...(ii)}$$

$$\therefore \quad \theta = \tan^{-1} = \frac{m_1 - m_2}{1 + m_1 m_2}$$

Notes. 1. *If ϕ is the other angle between the two given lines then*

$$\phi = \pi - 0$$

or
$$\tan \phi = \tan (\pi - \theta) = - \tan \theta$$

$$= \frac{m_1 - m_2}{1 + m_1 m_2}. \qquad \text{...(iii)}$$

$\therefore$ Combining the results (ii) and (iii) we have

$$\tan \theta = \pm \frac{m_1 - m_2}{1 + m_1 m_2}$$

which is the complete *angle formula* between two give lines.

2. *If* $\frac{m_1 - m_2}{1 + m_1 m_2}$ *is +ve, then* $\tan^{-1} \frac{m_1 - m_2}{1 + m_1 m_2}$ *is the acute angle between the lines and if* $\frac{m_1 - m_2}{1 + m_1 m_2}$ *is negative, then* $\tan^{-1} \frac{m_1 - m_2}{1 + m_1 m_2}$ *is the obtuse angle between the given lines.*

3. *To numerical examples, if the angle between two lines is required, we discard the negative sign and find the acute angle between the lines.*

Cor. *To find the condition that the two straight lines $y = m_1 x + c_1$ and $y = m_2 x + c_2$ may be*

(i) parallel,

(ii) perpendicular.

Proof:

The given lines are

$$y = m_1 x + c_1$$

and $y = m_2 x + c_2$.

If θ is the angle between the lines, then

$$\tan \theta = \frac{m_1 - m_2}{1 + m_1 m_2}.$$

(i) If the lines are parallel, then $\theta = 0^\circ$

$$\therefore \quad \tan \theta = \tan 0^\circ = 0$$

$$\therefore \quad 0 = \frac{m_1 - m_2}{1 + m_1 m_2}$$

$\therefore \qquad m_1 - m_2 = 0$

or $\qquad m_1 = m_2$

which is the required condition of parallelism of two lines.

Hence, two given lines will be parallel if their slopes are equal and conversely.

(ii) If the lines are perpendicular to each other, then $\theta = 90^o$.

$\therefore \quad \tan\theta = \tan 90^o = \infty.$

$$\therefore \qquad \frac{m_1 - m_2}{1 + m_1 m_2} = \infty$$

$$\Rightarrow \qquad \frac{m_1 - m_2}{1 + m_1 m_2} = \frac{1}{\infty} = 0$$

$$\Rightarrow \qquad 1 + m_1 m_2 = 0$$

$$\Rightarrow \qquad m_1 m_2 = -1$$

which is the required condition for two lines to be perpendicular.

Hence, two given lines are perpendicular if the product of their slopes is equal to –1 and conversely. i.e.

$$m_1 \times m_2 = -1$$

To find the angle between the straight lines

$$a_1x + b_1y + c_1 = 0$$

and $a_2x + b_2y + c_2 = 0.$

Proof:

Let m_1 and m_2 be the slopes of the two given lines, then we have

$$a_1x + b_1y + c_1 = 0$$

and $a_2x + b_2y + c_2 = 0.$

Then we have $m_1 = \frac{a_1}{b_1}$ and $m_2 = \frac{a_2}{b_2}$.

If θ is the angle between the lines, then

$$\tan\theta = \frac{m_1 - m_2}{1 + m_1 m_2} = \frac{\left(\frac{a_2}{b_2}\right) - \left(\frac{a_1}{b_1}\right)}{1 + \left(-\frac{a_1}{b_1}\right)\left(-\frac{a_2}{b_2}\right)}$$

$$\Rightarrow \frac{\frac{a_2}{b_2}-\frac{a_1}{b_1}}{1+\frac{a_1a_2}{b_1b_2}} = \frac{a_2b_1-b_2a_1}{a_1a_2+b_1b_2}$$

$$\therefore \quad \theta = \tan^{-1}\left(\frac{a_2b_1-b_2a_1}{a_1a_2+b_1b_2}\right)$$

Cor. : *Conditions for the lines*

$$a_1x + b_1y + c_1 = 0$$

and $a_2x + b_2y + c_2 = 0$

may be (i) parallel, (ii) perpendicular. are given as

(i) If the lines are parallel, then $\theta = 0^\circ$

$\therefore \quad \tan\theta = \tan 0^\circ = 0$

$$\therefore \quad 0 = \frac{a_2b_1-b_2a_1}{a_1a_2+b_1b_2}$$

$$\Rightarrow \quad a_2b_1 - a_1b_2 = 0$$

$$\Rightarrow \quad \frac{a_1}{a_2} = \frac{b_1}{b_2}$$

(ii) If the lines are perpendicular, then $\theta = 90^\circ$

$\therefore \quad \tan\theta = \tan 0^\circ = \infty$

$$\therefore \quad \frac{a_2b_1-a_1b_2}{a_2b_1+a_1b_2} = \infty$$

$$\Rightarrow \quad \frac{a_1a_2+b_1b_2}{a_2b_1-a_1b_2} = \frac{1}{\infty} = 0$$

$$\Rightarrow \quad a_1a_2 + b_1b_2 = 0.$$

SYSTEM OF THE STRAIGHT LINES

In this section, we shall give the general formula for writing the equation of straight line (i) parallel to a given straight line, (ii) perpendicular to given straight line.

(i) Lines perpendicular to ax + by + c = 0

The slope of this line is $\left(\frac{-a}{b}\right)$, therefore the slope of the line which are perpendicular to the line will be $\left(\frac{-1}{-a/b}\right) = \frac{b}{a}$. Hence the equation of line having b/a as slope can be written as

$$y = \left(\frac{b}{a}\right) x + \lambda$$

$$\Rightarrow \qquad bx - ay + \lambda = 0.$$

where λ is any arbitrary constant.

Hence the equation of a line perpendicular to a given line is obtained by (i) interchanging the co-efficients of x and y, (ii) changing the sign of one of them, and (iii) adding the constant term whose value can be obtained under the given condition.

(ii) Lines parallel to $ax + by + c = 0$.

Since the slopes of parallel lines are equal, the slope of the new line parallel to given line will be $(-a/b)$. Equation of the line with this slope is

$$y = \left(\frac{-a}{b}\right) x + \lambda$$

$$\Rightarrow \qquad ax + by + \lambda = 0,$$

where λ is any arbitrary constant.

Hence, the equation of a straight line parallel to a given straight line is obtained by changing the constant term whose value can be obtained subject to given conditions.

Example:

Find the area of the triangle formed. By the lines whose equations are $y = m_1x + c_1$, $y = m_2x + c_2$ *and* $x = 0$.

Solution:

The equation of the sides of a triangle are

$$y = m_1x + c_1 \qquad \text{...(i)}$$

$$y = m_2x + c_2 \qquad \text{...(ii)}$$

$$x = 0. \qquad \text{(iii)}$$

The point of intersection of (i) and (ii) is obtained by solving these two equations simultaneously. This gives

$$\left(\frac{c_2 - c_1}{m_1 - m_2}, \frac{c_2 m_1 - c_1 m_2}{m_1 - m_2}\right).$$

Similarly, $(0, c_1)$ is the point of intersection (i) and (iii) and $(0, c_2)$ is the point of intersection of (ii) and (iii). Hence the required area of the triangle is

$$\frac{1}{2}\begin{vmatrix} 0 & c_1 & 1 \\ 0 & c_2 & 1 \\ \frac{c_2 - c_1}{m_1 - m_2} & \frac{c_2 m_1 - c_1 m_2}{m_1 - m_2} & 1 \end{vmatrix}$$

$$= \frac{1}{2}\left(\frac{c_2 - c_2}{m_1 - m_2}\right)(c_1 - c_2)$$

$$= \frac{1}{2}\frac{(c_2 - c_2)^2}{m_1 - m_2}$$

SOLVED EXAMPLES

Example 1:

Find what straight lines are represented by the following equation and determine the angle between them

$$33x^2 - 71xy - 14y^2 = 0.$$

Solution:

$$33x^2 - 71y - 14y^2 = 0.$$

Factorizing, $(11x + 2y)(3x - 7y) = 0$.

Hence the equations are $11x + 2y = 0$

and $3x - 7y = 0$.

$$\tan\theta = \frac{2 - \sqrt{\left\{\left(-\frac{71}{2}\right)^2 + 33.14\right\}}}{33 - 14} = \frac{83}{19} \quad \therefore \ \theta = \tan^{-1}\left(\frac{83}{19}\right).$$

Example 2:

Find what straight lines are represented by the following equation and determine the angles between them

$$x^3 - 6x^2 - 11x - 6 = 0.$$

Solution:

By remainder theorem, the factors are

$$(x - 1)(x - 2)(x - 3) = 0.$$

Hence, the lines represented will be

$$x - 1 = 0,$$

$$x - 2 = 0$$

and $x - 3 = 0$,

and clearly all the lines are parallel to y-axis. **Ans.**

Example 3:

Prove that the angle between the straight lines joining the origin to the intersection of the straight line y = 3x + 2 with the curve

$$x2 + 2xy + 3y^2 + 4x + 8y - 11 = 0 \text{ is } \tan^{-1}\frac{2\sqrt{2}}{3}.$$

Solution:

The equation of the curve is

$$x^2 + 2xy + 3y^2 + 4x + 8y - 11 = 0$$

The line is given as

$$y = 3x + 2 \text{ or } \frac{y}{2} - \frac{3x}{2} = 1. \quad \text{..(1)}$$

The curve is given as $x^2 + 2xy + 3y^2 + 4x + 8y - 11 = 0$ ...(2)

Making (2) homogeneous with the help of (1), we get

$$x^2 + 2xy + 3y^2 + 4x\left(\frac{y}{2} - \frac{3x}{2}\right) + 8y\left(\frac{y}{2} - \frac{3x}{2}\right) - 11\left(\frac{y}{2} - \frac{3x}{2}\right)^2 = 0$$

$$\Rightarrow \quad x^2\left(1 - 6 - \frac{99}{4}\right) + y^2\left(3 + 4 - \frac{11}{4}\right) + xy\left(2 + 2 - 12 + \frac{33}{2}\right) = 0$$

$$\Rightarrow \quad -\frac{119}{4}x^2 + \frac{17}{4}y^2 + \frac{17}{2}xy = 0$$

$$\Rightarrow \quad 7x^2 - 2xy - y^2 = 0.$$

If ϕ be the angle between them, then

$$\tan\theta = \frac{\pm 2\sqrt{\{(-1)^2 - 7(-1)\}}}{7 + (-1)} = \frac{2\sqrt{8}}{6} = \frac{2\sqrt{2}}{3}.$$

$$\therefore \quad \theta = \tan^{-1}\left(\frac{2\sqrt{2}}{3}\right).$$

Example 4:

What loci are represented by the equation $xy - ay = 0$.

Solution:

We have $xy - ay = 0 \quad \Rightarrow y(x - a) = 0$

So the given line represents 2 straight lines as

$y = 0$, i.e., x-axis and $(x - a) = 0$,

i.e., parallel to y-axis at a distance a.

Example 5:

What loci are represented by the equation $x^3 - x^2 - x + 1 = 0$.

Solution:

The given equation is

$$x^3 - x^2 - x + 1 = 0$$

$$\Rightarrow \quad (x - 1)(x - 1)(x + 1) = 0 \text{ (on factorizing).}$$

Hence the equation represents 3 straight lines, given by

$x - 1 = 0$, $x - 1 = 0$ and $x + 1 = 0$.

the first two coinciding and all are parallel to y-axis at a distance ± 1.

Example 6:

What loci are represented by the equation $x^3 - xy^2 = 0$.

Solution:

The given equation is

$x^3 - xy^2 = 0$ or $x(x - y)(x + y) = 0$.

Hence the equation represents 3 straight lines as

$x = 0$, i.e., y-axis $x - y = 0$ and $x + y = 0$,

last two being inclined at an angle of 45° and 135° to x-axis and passing through origin.

Example 7:

Show that the straight lines joining the origin to the other two points of intersection of the curves whose equations are $ax^2 + 2hxy + by^2 + 2gx = 0$ *and* $a'x^2 + 2h'xy + b'y^2 + 2g'x = 0$ *will be at right angles if* $g(a' + b') - g'(a + b) = 0$.

Solution:

The two curves are given by

$$ax^2 + 2hxy + by^2 + 2gx = 0, \qquad ...(1)$$

$$a'x^2 + 2h'xy + b'y^2 + 2g'x = 0. \qquad ...(2)$$

To get the combined equation of the straight lines joining the point of intersection of (1) and (2) to the origin, we make (2) homogeneous with the help of (1), i.e., we eliminate the terms in first degree from (1) and (2). Hence multiplying (1) by g' and (2) by g and subtracting, we get

$$(g'ax^2 + 2g'hxy + g'by^2 + 2gg'x) - (ga'x^2 + 2h'gxy + gb'y^2 + 2gg'x) = 0$$

$$\Rightarrow x^2 (ag' - ga') + 2 (g'h - gh') xy + (g'b - gb') y^2 = 0, \qquad ...(3)$$

which is the required equation.

If the lines represented by (3) be at right angles, then

$$(ag' - ga') + (g'b - gb') = 0$$

$$\Rightarrow \quad g' (a + b) - g (a' + b') = 0$$

$$\Rightarrow \quad g (a' + b') - g' (a + b) = 0.$$

Example 8:

Show that the straight lines $(A^2 - 3B^2) x^2 + 8ABxy + (B^2 - 3A^2) y^2 = 0$, form with the line $Ax + By + C = 0$ an equilateral triangle whose area is

$C^2/\sqrt{3}(A^2 + B^2)$.

Solution:

The given equation is

$$(A^2 - 3B^2)(x^2 + 8ABxy + (B^2 - 3A^2) y^2 = 0.$$

As it is homogenous equation in second degree, so represents 2 straight lines passing through origin.

Factorizing, we get

$$\{(\sqrt{3}A - B) y - (A + \sqrt{3}B)x\} \{(\sqrt{3}A + B) y + (A - \sqrt{3}B) x\} = 0.$$

Hence the lines represented by the given equation will be

$$(\sqrt{3}A - B) y - (A + \sqrt{3}B) x = 0 \qquad ...(1)$$

and $(\sqrt{3}A + B) y + (A - \sqrt{3}B) x = 0 \qquad ...(2)$

The third line is given as

$$Ax + By + C = 0. \qquad ...(3)$$

If α be the angle between (1) and (3), β be the angle between (2) and (3), then

$$\tan\alpha = \pm\left\{\frac{\left(\frac{A+\sqrt{3}B}{\sqrt{3}A-B}\right)-\left(-\frac{A}{B}\right)}{1+\left(\frac{A+\sqrt{3}B}{\sqrt{3}A-B}\right)\cdot\left(-\frac{A}{B}\right)}\right\} = \pm\frac{\sqrt{3}\left(A^2+B^2\right)}{A^2+B^2} = \sqrt{3}$$

$\therefore$ $\alpha = 60^\circ$ (taking acute angle).

Similarly,

$$\tan\beta = \pm\left\{\frac{\left(\frac{A-\sqrt{3}B}{\sqrt{3}A+B}\right)-\left(-\frac{A}{B}\right)}{1+\left(\frac{A-\sqrt{3}B}{\sqrt{3}A+B}\right)\cdot\left(\frac{A}{B}\right)}\right\} = \pm\frac{\sqrt{3}\left(A^2+B^2\right)}{A^2+B^2} = \pm\sqrt{3}$$

$\therefore$ $\beta = 60^\circ$.

As two angles of the triangle are 60° each, the triangle is an equilateral one.

If p be the altitude of an equilateral triangle, its area may be given by

$$\text{Area} = \frac{p^2}{\sqrt{3}}.$$

The perpendicular distance from origin (vertex of the Δ) on opposite side $Ax + By + C = 0$ is

$$\frac{C}{\sqrt{(A^2+B^2)}}.$$

Hence the area of the triangle $= \dfrac{C^2}{\sqrt{3}\left(A^2+B^2\right)}$. unit **Ans.**

Example 9:

Prove that the equation $y^3 - x^3 + 3xy\,(y-x) = 0$, represents three straight lines equally inclined to one another.

Solution:

The given equation is

$$y^3 - x^3 + 3xy\,(y - x) = 0$$

$\Rightarrow$ $(y - x)(y^2 + xy + x^2) + 3xy\,(y - x) = 0.$

$\Rightarrow$ $(y - x)(y^2 + xy + x^2 + 3xy) = 0.$

$\Rightarrow$ $(y - x)\,[y + (2 - \sqrt{(3)}\,x]\,(y + (2 + \sqrt{(3)}\,x] = 0.$

Hence $y - x = 0$ ...(1)

$y + (2 - \sqrt{3})\,x = 0$ ...(2)

and $$y + (2 + \sqrt{3})\, x = 0 \qquad ...(3)$$

will be the three lines represented by the given equations.

If α_1, α_2 and α_3 be angles between (1), (2); (2), (3) and (1), (3), respectively, then we have

$$\tan \alpha_1 = \pm \frac{1+(2-\sqrt{3})}{1-1\,(2-\sqrt{3})} = \pm \frac{3-\sqrt{3}}{\sqrt{3}-1} = \pm \frac{\sqrt{3}\,(\sqrt{3}-1)}{\sqrt{3}-1} = \pm\sqrt{3},$$

$$\tan \alpha_2 = \pm \frac{-(2-\sqrt{3})+(2+\sqrt{3})}{1+(2-\sqrt{3})\,(2+\sqrt{3})} = \pm \frac{2\sqrt{3}}{2} = \pm\sqrt{3},$$

$$\tan \alpha_2 = \pm \frac{1+(2+\sqrt{3})}{1-(2+\sqrt{3})} = \pm \frac{3+\sqrt{3}}{-\sqrt{3}-1} = \pm \frac{3\,(\sqrt{3}+1)}{\sqrt{(3+1)}} = \pm\sqrt{3}.$$

Hence $$\alpha_1 = \alpha_2 = \alpha_3.$$

Example 10:

The equations to a pair of opposite sides of a parallelogram are $x^2 - 7x + 6 = 0$ and $y^2 - 14y + 40 = 0$; find the equations to its diagonals.

Solution:

The equations of one set of opposite sides is $x^2 - 7x + 6 = 0$ or $(x - 6)(x - 1) = 0$.

Hence the equations of the opposite sides say AB and CD respectively are

$$x - 6 = 2 \qquad ...(1)$$

and $x - 1 = 0$ ...(2)

The equation of the other set of opposite sides is

$$y^2 - 14y + 40 = 0$$

or $$(y - 4)(y - 4)(y - 10) = 0.$$

So the equations of other two sides say AD and BC respectively are

$$y - 4 = 0 \qquad ...(3)$$

and $y - 10 = 0$. ...(4)

Solving (1) and (3), the co-ordinates of A are (6, 4).

Solving (1) and (4), the co-ordinates of B are (6, 10).

Solving (2) and (4), the co-ordinates of C are (1, 10).

Solving (2) and (3), the co-ordinates of D are (1, 4).

Equation of AC is $y - 4 = \frac{10-4}{1-6}(x-6)$

$\Rightarrow \quad 5y + 6x = 56.$

Equation of BD is $y - 10 = \frac{4-10}{1-6}(x-6)$

$\Rightarrow \quad 5y - 6x = 14.$

Example 11:

What relations must hold between the coefficients of the equations (i) $ax^2 + by^2 + cx + cy = 0$. (ii) $ay^2 + bxy + dy + cx = 0$, so that each of them may represent a pair of straight lines ?

Solution:

The given equation is

(i) $ax^2 + by^2 + cx + cy = 0$

If it represents 2 straight lines, D must be zero.

Hence $\Delta = a.b.0 + 2.\frac{c}{2}.\frac{c}{2}0 - a.\left(\frac{c}{2}\right)^2 - b\left(\frac{c}{2}\right)^2 - 0 = 0$

or $\quad -\frac{ac^2}{4} - \frac{bc^2}{4} = 0$ or $c^2(a+b) = 0$.

So either $c = 0$ or $(a + b) = 0$ or $c(a + b) = 0$.

(ii) $D = 0.a.0 + 2.\frac{d}{2}.\frac{c}{2}.\frac{b}{2} - 0\left(\frac{d}{2}\right)^2 - a\left(\frac{c}{2}\right)^2 - 0\left(\frac{b}{2}\right)^2 = 0$

or $\quad \frac{bdc}{4} - \frac{ac^2}{4} = 0$ or $c(bd - ac) = 0$.

Hence, either $c = 0$ or $bd - ac = 0$.

Example 12:

What loci are represented by the equation $x^2 - y^2 = 0$.

Solution:

The given equation is

$x^2 - y^2 = 0$ or $(x - y)(x + y) = 0$.

Hence the equation represents 2 straight lines

$x - y = 0$ and $x + y = 0$.

Example 13:

What loci are represented by the equation $x^2 - xy = 0$.

Solution:

The given equation is

$$x^2 - xy = 0 \text{ or } x(x - y) = 0.$$

So the given equation represents two straight lines as $x = 0$, i.e., y-axis and $(x - y) = 0$.

Example 14:

What loci are represented by the equation $x^3 + y^3 = 0$.

Solution:

Hence, the equation represents three straight lines first being $x + y = 0$, i.e., passing through origin and inclined at an angle of 135° to x-axis, the other two being imaginary as $x^2 - xy + y^2 = 0$ has no real roots, but it represents two straight lines, passing through origin, being homogeneous in second degree.

Example 15:

What loci are represented by the equation $x^2 + y^2 = 0$.

Solution:

The given equation is

$$x^2 + y^2 = 0.$$

As the expression on L.H.S. cannot be factorized involving real factors and is a homogeneous equation in second degree, so represents 2 imaginary lines passing through origin.

In real variables, the relation is only possible when both $x = 0$ and $y = 0$ are satisfied; so it represents a point, i.e., origin.

Example 16:

What loci are represented by the equation $x^2y = 0$.

Solution:

So the equation represents three straight lines, as $x = 0$ and $y = 0$, the first two coinciding representing axis of y and the third one representing axis.

Example 17:

What loci are represented by the equation $(x^2 - 1)(y^2 - 4) = 0$.

Solution:

$$(x^2 - 1)(y^2 - 4) = 0$$

$$\Rightarrow \qquad (x + 1)(x - 1)(y + 2)(y - 2) = 0$$

Hence, the given equation represents four st. lines i.e. $x = \pm 1$ (Par to axis) and $y = \pm 2$ (Parallel to x-axis)

Example 18:

Find the equation to the pair of straight lines joining the origin to the intersections of the straight line $y = mx + c$ and the curve

$$x^2 + y^2 = a^2.$$

Prove that they are at right angles, if $2c^2 = a^2 (1 + m^2)$.

Solution:

Straight line is given as $y = mx + c$ or $\dfrac{y}{c} - \dfrac{mx}{c} = 1.$ ...(1)

The curve is given as $x^2 + y^2 = a^2$ or $x^2 + y^2 - a^2 = 0.$ ...(2)

Making (2) homogeneous with the help of (1), we get

$$\Rightarrow \qquad x^2 + y^2 - a^2 \left(\frac{y}{c} - \frac{mx}{c^2}\right)^2 = 0$$

$$x^2 + y^2 - a^2 \left[\frac{y^2}{c^2} + \frac{m^2x^2}{c^2} - \frac{2myx}{c^2}\right] = 0$$

$$\Rightarrow \qquad c^2x^2 + c^2y^2 - a^2y^2 - a^2m^2x^2 + 2a^2myx = 0.$$

$$\Rightarrow \qquad x^2 (c^2 - a^2m^2) + y^2 (c^2 - a^2) + 2a^2myx = 0.$$

...(3)

This equation represents the two lines joining the meeting point of (1) and (2) with the origin.

If the lines represented by (3) are at right angles, then

$$(c^2 - a^2m^2) + (c^2 - a^2) = 0$$

[as the sum of the coefficients of x and y^2 must be zero]

$$2c^2 = a^2 (1 + m^2).$$

Example 19:

Prove that the straight lines joining the origin to the points of intersection of the straight line $kx + hy = 2hk$ with the curve

$$(x - h)^2 (y - k)^2 = c^2 \textit{ are at right angles if, } h^2 + k^2 = c^2.$$

Solution:

Then given is $kx + hy = 2hk$ or $\dfrac{x}{2h} + \dfrac{y}{2k} = 1.$ ...(1)

The given curve is $(x - h)^2 + (y - k)^2 = c^2$

$\Rightarrow$ $x^2 + y^2 - 2hx - 2ky + h^2 + k^2 - c^2 = 0.$ ...(2)

Making (2) homogeneous with the help of (1), we get

$$x^2 + y^2 - 2hx\left(\frac{x}{2h}+\frac{y}{2k}\right) - 2ky\left(\frac{x}{2h}+\frac{y}{2k}\right) + \left(h^2+k^2-c^2\right)\left(\frac{x}{2h}+\frac{y}{2k}\right)^2 = 0.$$

...(3)

Coefficient of x^2 in (3) is $= \dfrac{h^2+k^2-c^2}{4h^2}$

Coefficient of y^2 in (3) is $= \dfrac{h^2+k^2-c^2}{4k^2}$.

If the lines represented by (3) are perpendicular to each other, then

$$\frac{h^2+k^2-c^2}{4h^2}+\frac{h^2+k^2-c^2}{4k^2}=0$$

$$\Rightarrow \qquad \left(h^2+k^2-c^2\right)\left(\frac{1}{4h^2}+\frac{1}{4k^2}\right)=0.$$

As $\left(\dfrac{1}{4h^2}+\dfrac{1}{4k^2}\right)$ cannot be zero, being sum of two squares, hence.

Example 20:

Show that the equation $bx^2 - 2xy + ay^2 = 0$ represents a pair of straight lines which are at right angles to the pair given by the equation $ax^2 + 2hxy + by^2 = 0$.

Solution :

First equation is given as, $bx^2 - 2hxy + ay^2 = 0$ ...(1)

Let the equation represented by (1) be

$y = m_1x$ ...(2)

and $y = m_2x$. ...(3)

Then $(y - m_1x)(y - m_2x) = 0$ must be same as (1)

or $y^2 - (m_1 + m_2)xy + m_1m_2x^2 = 0$ must be identical with $bx^2 - 2hxy + ay^2 = 0$.

Comparing the coefficients, $\dfrac{1}{a} = \dfrac{-(m_1+m_2)}{-2h} = \dfrac{m_1\,m_2}{b}$.

Hence $$m_1 m_2 = \frac{b}{a} \qquad ...(4)$$

$$m_1 + m_2 = \frac{2h}{a}. \qquad ...(5)$$

The line perpendicular to (2) is $y = -\frac{1}{m^2} x$ or $m_2 y + x = 0$.

Combined equation of these two is $(m_1 y + x)(m_2 y + x) = 0$

$$\Rightarrow \quad m_1 m_2 y_2 + (m_1 + m_2) xy + x^2 = 0.$$

Substituting the values from (4) and (5), we get

$$\frac{b}{a} y^2 + \frac{2h}{a} xy + x^2 = 0$$

$$\Rightarrow \quad by^2 + 2hxy + ax^2 = 0$$

which is same as the given second equation.

Example 21:

If pairs of straight lines $x^2 - 2pxy - y^2 = 0$ and $x^2 - 2qxy - y^2 = 0$ be such that each pair bisects the angles between the other pair, prove the $pq = 1$.

Solution :

The given equation is $x^2 - 2pxy - y^2 = 0$. ...(1)

As the equation of the pairs of the lines bisecting the angles between the lines $ax^2 + 2hxy + by^2 = 0$ is $\frac{x^2 - y^2}{a - b} = \frac{xy}{h}$.

Hence the equation of the lines bisecting the angles between (1) will be

$$\frac{x^2 - y^2}{1 - (-1)} = \frac{xy}{-p}.$$

or $$px^2 + 2xy - py^2 = 0. \qquad ...(2)$$

The given equation of the bisector is $x^2 - 2qxy - y^2 = 0$. ...(3)

As (2) and (3) represent the same straight lines, comparing the coefficients,

$$\frac{p}{1} = \frac{2}{-2q} \text{ or } pq = -1.$$

Example 22:

Prove that the pair of lines $a^2x^2 + 2h(a + b)xy + b^2y^2 = 0$ is equally inclined to the pair $ax^2 + 2hxy + by^2 = 0$.

Solution:

If each line of the pair of straight lines be equally inclined to the lines of the other pair of straight lines, they will have a common bisector.

The first equation is given as

$$a^2x^2 + 2h(a + b)xy + b^2y^2 = 0. \quad ...(1)$$

The equation of the bisector will be

$$\frac{x^2 - y^2}{a^2 - b^2} = \frac{xy}{h(a+b)} \text{ or } \frac{x^2 - y^2}{a - b} = \frac{xy}{h}. \quad ...(2)$$

The other equation is given as $ax^2 + 2hxy + by^2 = 0$. ...(3)

The equation of bisector clearly $\frac{x^2 - y^2}{a - b} = \frac{xy}{h}$ which is same as (2), i.e., bisector of (1).

Example 23:

Prove that the equation to the bisectors of the angle between the straight lines $ax^2 + 2hxy + by^2 = 0$ is $h(x^2 - y^2) + (b - a)xy = (ax^2 - by^2)\cos\omega$, the axes being inclined at an angle ω.

Solution:

Let X'OX and YOY' be the axes inclined at an angle ω.CD and AB be two straight lines represented by given equation $ax^2 + 2hxy + by^2 = 0$ inclined at angles θ_1 and θ_2 to x-axis respectively. Let PQ and RS be the bisectors of the angles between the lines AB and CD. If PQ is inclined at an angle θ to x-axis, RS will be inclined at $(\pi/2 + \theta)$.

Again $\theta + \theta_1 + \frac{1}{2}(\theta_1 - \theta_2) = \frac{\theta_1 + \theta_2}{2}$

or $\qquad 2\theta = (\theta_1 + \theta_2) \quad ...(1)$

Let CD be represented by $y = m_1x$ and AB by $y = m_2x$. Hence the combined equation will be $(y - m_1x)(y - m_2x) = 0$.

Comparing with given equation, we get

$$m_1 + m_2 = -\frac{2h}{b}$$

and $m_1 m_2 = \frac{a}{b}$.

Now $\tan 2\theta = \tan(\theta_1 + \theta_2) = \frac{\tan\theta_1 + \tan\theta_2}{1 - \tan\theta_1 \tan\theta_2}$

$$= \frac{\dfrac{m_1 \sin \omega}{1+m_1 \cos \omega} + \dfrac{m_2 \sin \omega}{1+m_2 \cos \omega}}{1 - \dfrac{m_1 \sin \omega}{1+m_1 \cos \omega} \cdot \dfrac{m_2 \sin \omega}{1+m_2 \cos \omega}}$$

$$\therefore \left[\begin{array}{l}\text{if the line} = m_1 x \text{ makes an angle } \theta_1 \text{ from } x-\text{axis} \\ \tan q_1 = \dfrac{m_1 \sin \omega}{1+m_1 \cos \omega}\end{array}\right]$$

$$= \frac{2m_1 m_2 \sin \omega \cos \omega + (m_1 + m_2) \sin \omega}{1+(m_1+m_2) \cos \omega - m_1 m_2 (\sin^2 \omega - \cos^2 \omega)}$$

$$\therefore \tan 2\theta = \frac{2a \sin \omega \cos \omega - 2h \sin \omega}{b - 2h \cos \omega - a(\sin^2 \omega - \cos^2 \omega)}$$

$$\left(\therefore \quad m_1 m_2 = -\frac{2h}{b}, m_1 m_2 = \frac{a}{b}\right) \qquad ...(2)$$

Again let PQ be y = mx. ...(3)

As PQ is inclined at an angle of θ to x-axis $\tan \theta = \dfrac{m \sin \omega}{1 + m \cos \omega}$

$$\therefore \quad \tan 2\theta = \frac{2 \tan \theta}{1 - \tan^2 \theta} = \frac{2 m \sin \omega / (1 + m \cos \omega)}{1 - \{m \sin \omega / (1 + m \cos \omega)\}^2}.$$

$$= \frac{2\dfrac{y}{x} \sin \omega \Big/ \left(1 + \dfrac{y}{x}.\cos \omega\right)}{1 - \left\{\dfrac{y}{x} \sin \omega \Big/ \left(1 + \dfrac{y}{x} \cos \omega\right)\right\}} \quad \left(\therefore \quad m = \frac{y}{x} \text{ by (3)}\right)$$

$$= \frac{2y \sin \omega \, (x + y \cos \omega)}{(x + y \cos \omega)^2 - (y \sin \omega)^2} \qquad ...(4)$$

Equating R.H.S. of (2) and (4), as L.H.S. are same, we get

$$\frac{2a \sin \omega \cos \omega - 2h \sin \omega}{b - 2h \cos \omega - a(\sin^2 \omega - \cos^2 \omega)} = \frac{2y \sin \omega \, (x + y \cos \omega)}{(x + y \cos \omega)^2 - (y \sin \omega)^2}$$

Simplifying we get

$$h(x^2 - y^2) - (b - a)xy = (ax^2 - by^2) \cos \omega.$$

Example 24:

If one of the straight lines given by the equation $ax^2 + 2hxy + by^2 = 0$, *coincide with one of those given by $a'x^2 + 2h'xy + b'y^2 = 0$, and the other lines represented by them the perpendicular, prove that*

$$\frac{ha'b'}{b'-a'} = \frac{h'ab}{b-a} = \frac{1}{2}\sqrt{-aa'bb'}.$$

Solution:

The equations are given as

$$ax^2 + 2hxy + by^2 = 0 \qquad ...(1)$$

and $a'x^2 + 2h'xy + b'y^2 = 0$. ...(2)

Let the equations represented by (2) be

$$y = m_1x \qquad ...(3)$$

and $y = m_2x$. ...(4)

Then combined equation of (3) and (4) will be given as

$$(y - m_1x)(y - m_2x) = 0$$

or $$y^2 - (m_1 + m_2)xy + m_1m_2x^2 = 0. \qquad ...(5)$$

As (a) and (5) represent the same equation hence comparing the coefficients

$$\frac{1}{b} = -\frac{m_1 + m_2}{2h} = \frac{m_1 m_2}{a}.$$

$$\therefore \qquad m_1 + m_2 = -\frac{2h}{b} \qquad ...(6)$$

and $m_1 m_2 = \dfrac{a}{b}$. ...(7)

As (2) represents a straight line coinciding with (3) and other perpendicular to (4), we must have (2) similar to $(y - m_1x)(m_2y + x) = 0$.

Comparing with (2), we get

$$\frac{b'}{m_2} = \frac{2h'}{1 - m_1 m_2} = \frac{a'}{-m_1} \text{ or } \frac{m_1}{m_2} = -\frac{a'}{b'} \qquad ...(8)$$

and $$\frac{m_1 m_2 - 1}{m_2} = -\frac{2h'}{b'}$$

$$\Rightarrow \qquad m_1 - \frac{1}{m_2} = -\frac{2h'}{b'} \qquad ...(9)$$

Solving (6) and (8), we get $m_1 = \dfrac{-2a'h}{b(a'-b')}$ and $m_2 = \dfrac{2b'h}{b(a'-b')}$

Substituting in (7), we get $\dfrac{-2a'h}{b(a'-b')}\cdot\dfrac{2b'h}{b(a'-b')}=\dfrac{a}{b}$

$$\Rightarrow \quad \frac{4h^2a'b'}{(a'-b')^2}=-ab$$

$$\Rightarrow \quad \frac{4h^2a'^2\,b'^2}{(b'-a')^2}=-aba'b'$$

$$\Rightarrow \quad \frac{ha'b'}{b'-a'}=\frac{1}{2}\sqrt{(-aba'b')}\,.$$

Similarly by (7) and (8), $m_1^2=-\dfrac{aa'}{bb'}$ and $\dfrac{1}{m_2^2}=\left(-\dfrac{a'b}{ab'}\right)$. ...(10)

Putting in (9), $\sqrt{\left(-\dfrac{aa'}{bb'}\right)}-\sqrt{-\left(\dfrac{a'b}{ab'}\right)}=-\dfrac{2h'}{b'}$

$$\Rightarrow \quad \sqrt{\left(-\frac{a'}{b'}\right)}\left[\frac{\sqrt{a}}{\sqrt{b}}-\frac{\sqrt{b}}{\sqrt{a}}\right]=-\frac{2h'}{b'}$$

$$\Rightarrow \quad \sqrt{\left(-\frac{a'}{b'}\right)}\left[\frac{a-b}{\sqrt{(ab)}}\right]=-\frac{2h'}{b'}$$

$$\Rightarrow \quad \sqrt{(-a'abb)}.(a-b)=-2h'ab$$

$$\Rightarrow \quad \frac{h'ab}{b-a}=\frac{1}{2}\sqrt{(-aa'bb')} \qquad ...(11)$$

By (10) and (11), we get the required result.

Example 25:

Show that the distance between the points of intersection of the straight line $x\cos\alpha+y\sin\alpha-p=0$, with the straight lines $ax^2+2hxy+by^2=0$ is $\dfrac{2p\sqrt{h^2-ab}}{b\cos^2\alpha-2h\cos\alpha\sin\alpha+\alpha\sin^2\alpha}$

Deduce the area of the triangle formed by them.

Solution:

The equations are given as

$$x\cos\alpha+y\sin\alpha-p=0 \qquad ...(1)$$

and $\quad ax^2 + 2hxy + by^2 = 0$ ...(2)

To above them, putting the value of x from (1) in (2), we get

$$a.\left(\frac{p - y \sin \alpha}{\cos \alpha}\right) + 2h.\frac{p - y \sin \alpha}{\cos \alpha} y + by^2 = 0$$

$\Rightarrow \quad a (p - ya)^2 + 2h \cos \alpha (p - y \sin \alpha).y + by^2 \cos^2\alpha = 0$

$\Rightarrow \quad y^2 [a \sin^2\alpha - 2h \cos \alpha \sin \alpha + b \cos^2 \alpha]$

$\quad - 2y (ap \sin \alpha - ph \cos \alpha) + ap^2 = 0$

This being a quadratic equation will give us 2 values of y.

Let these values by y_1 and y_2, then

$$y_1 + y_2 = \frac{2p (a \sin \alpha - h \cos \alpha)}{a \sin^2 \alpha - 2h \cos \alpha \sin \alpha + b \cos^2 \alpha} \quad ...(3)$$

and $$y_1 y_2 = \frac{ap^2}{a \sin^2 \alpha - 2h \cos \alpha \sin \alpha + b \cos^2 \alpha} \quad ...(4)$$

If x_1 and x_2 be the corresponding values of abscissa for y_1 and y_2, then the points (x_1, y_1 and (x_2, y_2) will be the point of intersection of (1) and (2). So these will satisfy (1).

$\therefore \quad x_1 \cos \alpha + y_1 \sin \alpha - p = 0$

and $x_2 \cos \alpha + y_2 \sin \alpha - p = 0$.

Substracting, we get, $\cos \alpha (x_1 - x_2) + \sin \alpha (y_1 - y_2) = 0$. ...(5)

If d be the distance between the points (x_1, y_1) and (x_2, y_2), then $d^2 =$

$$\sqrt{[(x_1 - x_2)^2 + (y_1 - y_2)^2]} = \sqrt{[(y_1 - y_2)^2 \ \tan^2 \alpha + (y_1 - y_2)^2]}$$

$$\left\{\therefore \ (x_1 - x_2) = -\frac{(y_1 - y_2)}{\cos \alpha}\right\} \text{ by (5)}$$

$$= \sqrt{[(y_1 - y_2)^2 \ (\tan^2 \alpha + 1)]} = \sec \alpha \sqrt{[(y_1 + y_2)^2 4 \ (y_1 \ y_2)]}$$

$$= \sec \alpha \sqrt{\left[\left\{\frac{2p \ (a \sin \alpha - h \cos \alpha)}{a \sin^2 a - 2h \cos a \sin a + b \cos^2 a}\right. - \frac{4ap^2}{a \sin^2 \alpha - 2h \cos \alpha \sin \alpha + b \cos^2 \alpha}\right]}$$

$$= \frac{\sec \alpha \ 2p}{a \sin^2 \alpha - 2h \sin \alpha \cos \alpha + b \cos^2 \alpha}$$

$$\times [a^2 \sin^2 \alpha + h^2 \cos^2 \alpha - 2ah \sin \alpha \cos \alpha)]$$
$$- [a^2 \sin^2\alpha - 2ha \sin \alpha \cos \alpha + ab \cos^2 \alpha)].$$

$$= \frac{2p \sec \alpha \sqrt{\{\cos^2 \alpha \left(h^2 - ab\right)\}}}{a \sin^2 \alpha - 2h \sin \alpha \cos \alpha + b^2 \cos \alpha}$$

$$= \frac{2p \sec \alpha \cos \alpha \sqrt{\{\cos^2 \alpha \left(h^2 - ab\right)\}}}{a \sin^2 \alpha - 2h \sin \alpha \cos \alpha + a \sin^2 \alpha}$$

$$= \frac{2p \sqrt{\{\left(h^2 - ab\right)\}}}{b \cos^2 \alpha - 2h \sin \alpha \cos \alpha + a \sin^2 \alpha}$$

If the points of intersection be A and B, the line BA will be x cos α + y sin α – p = 0. The length of perpendicular on AB from the third vertex of the triangle which is origin, i.e., (0, 0) will be p (as the equation is in standard form in which p is the length of perpendicular from origin on the line).

So its area $= \frac{1}{2} \cdot \frac{p^2 \sqrt{\{\left(h^2 - ab\right)\}}}{b \cos^2 \alpha - 2h \cos \alpha \sin \alpha + a \sin^2 \alpha}$

$$= \frac{p^2 \sqrt{\{\left(h^2 - ab\right)\}}}{b \cos^2 \alpha - 2h \cos \alpha \sin \alpha + a \sin^2 \alpha}. \qquad \textbf{Ans.}$$

Example 26:

Prove that the general equation $ax^2 + 2hxy + by^2 + 2gx + 2fy + c = 0$, represents two parallel straight lines if $h^2 = ab$ and $bg^2 = af^2$ Prove also that the distance between them is $2\sqrt{\dfrac{g^2 - ac}{a\,(a + b)}}$.

Solution:

The given equation is

$$ax^2 + 2hxy + by^2 + 2gx + 2fy + c = 0. \qquad ...(1)$$

Let the two equations represented by (1) be respectively

$$\left(\sqrt{a}.x + \sqrt{b}.y + l\right) = 0. \qquad ...(2)$$

and $\left(\sqrt{a}.x + \sqrt{b}.y + m\right) = 0$.

Which is identical to (1) and as the coefficients of x^2 and y^2 are equal. Hence equating different coefficients

$$2h = 2\sqrt{(ab)} \qquad ...(4)$$

$$2g = \sqrt{a}\,(l+m) \qquad ...(5)$$

$$2f = \sqrt{b}\,(l+m) \qquad ...(6)$$

and $\quad c = l\,m \qquad ...(7)$

By (4), we get $\quad h^2 = ab$

Dividing (5) by (6), we get $\frac{g}{f} = \frac{\sqrt{a}}{\sqrt{b}}$ or $\frac{g^2}{f^2} = \frac{a}{b}$.

Hence $\quad bg^2 = af^2$.

Again, if p and p' be the lengths of perpendiculars from origin on (2) and (3), then the distance between them is p – p'.

$$\text{So } p - p' = \frac{l}{\sqrt{(a+b)}} - \frac{m}{\sqrt{(a+b)}} = \frac{(l-m)}{\sqrt{(a+b)}} = \sqrt{\left\{\frac{(l+m)^2 - 4lm}{a+b}\right\}}.$$

Example 27:

Prove that the product of the perpendiculars let fall from the point (x', y') upon the pair of straight lines

$$ax^2 + 2hxy + by^2 = 0 \text{ is}$$

$$\frac{ax^2 + 2hx'y' + by}{\sqrt{(a-b)^2 + 4h^2}}.$$

Solution:

Let the line $ax^2 + 2hxy + by^2 = 0$ represent the two straight lines

$$y - m_1x = 0 \qquad ...(1)$$

and $y - m_2x = 0$. $\qquad ...(2)$

$$m_1\, m_2 = \frac{a}{b} \qquad ...(4)$$

If p_1 and p_2 be the lengths of perpendiculars from (x_1, y_1) on (1) and (2) respectively, then we have

$$p_1\, p_2 = \frac{y_1 - m_1 x_1}{\sqrt{(1+m_1^2)}} \cdot \frac{y_1 - m_2 x_1}{\sqrt{(1+m_2^2)}}$$

$$= \frac{y_1^2 - x_1 y_1 (m_1 + m_2) + m_1\, m_2\, x_1^2}{y + m_1^2 + m_2^2 + (m_1\, m_2)^2}$$

$$= \frac{y_1^2 - x_1 y_1 (m_1 + m_2) + m_1 m_2 x_1^2}{1 + (m_1 + m_2)^2 - 2m_1 m_2 + (m_1 m_2)^2}$$

Putting the values of $m_1 + m_2$ and $m_1 m_2$ from (3) and (4),

$$p_1 p_2 = \frac{y_1^2 - \left(-\frac{2h}{b}\right) x_1 y_1 + \left(\frac{a}{b}\right) x_1^2}{\sqrt{\left(1 + -\frac{2h}{b}^2\right) - 2\frac{a}{b} + \left(\frac{a}{b}\right)^2}}$$

$$= \frac{by_1^2 + 2hx_1 y_1 + ax_1^2}{\sqrt{\{b^2 + 4h^2 - ab + a^2\}}} = \frac{ax_1^2 + 2hx.y_1 + by_1^2}{\sqrt{\{(a-b)^2 + 4h^2\}}}.$$

Example 28:

What loci are represented by the equation $(x - a)^2 - y^2 = 0$.

Solution:

The given equation is

$$(x - a)^2 - y^2 = 0.$$

Factorizing, we get $(x - a - y)(x - a + y) = 0$.

Hence the given equation represents two straight lines as $x - y - a = 0$ and $x + y - a = 0$. The first cutting an intercept of a and $-a$ from the axes and the second of a and a.

Example 29:

What loci are represented by the equation

$(x + y)^2 - c^2 = 0$ $(x + y - c)(x + y + c) = 0$.

Solution:

The given equation is

$$(x + y)^2 - c^2 = 0$$

or $$(x + y - c)(x + y + c) = 0.$$

Hence the equation represents 2 straight $x + y - c = 0$ and $x + y + c = 0$, first cutting the intercepts of c from both the axes second $-c$ from both the axes.

Example 30:

What loci are represented by the equation $r = a \sec(\theta - \alpha)$.

Solution:

The given equation is

$$r = a \sec(\theta - \alpha) \text{ or } r = \frac{a}{\cos(\theta - \alpha)} \text{ or } r\cos(\theta - \alpha) = a$$

$\Rightarrow$ $r(\cos\theta \cos\alpha + \sin\theta \sin\alpha) = a$

$\Rightarrow$ $r\cos\theta\cos\alpha + r\sin\theta\sin\alpha = a$

$\Rightarrow$ $x\cos\alpha + y\sin\alpha = a \quad (\because r\cos\theta = x, r\sin\theta = y).$

This is a standard form of a straight line. Hence the given equation represents a straight line.

Example 31:

What loci are represented by the equation

$$(x^2 - 1)^2 + (y^2 - 4)^2 = 0.$$

Solution:

The given equation is

$$(x^2 - 1)^2 + (y^2 - 4)^2 = 0.$$

As the sum of two squares cannot be equal to zero unless each of them is equal to zero, hence we must have both

$$x^2 - 1 = 0$$

and $y^4 - 4 = 0$

$\Rightarrow$ $(x - 1)(x + 1) = 0$

and $(y - 2)(y + 2) = 0.$

Hence, we get 4 different combinations, each representing a point. These points are (1, 2), (1, –2), (–1, 2) and (–1, 2).

Example 32:

What loci are represented by the equation

$$(y - mx - c)^2 + (y - mx - c')^2 = 0.$$

Solution:

The given equation is

$(y - mx + c)^2 + (y - mx - c')^2 = 0$

$$(y - mx - c)^2 = 0$$

or $\quad y - mx - c = 0 \quad$ ...(1)

and $(y - m'x - c)^2 = 0$

or $\quad y - m'x - c' = 0. \quad$ ...(2)

As both the equations (1) and (2) must be satisfied, the equation represents a point which is the point of intersection of (1) and (2). Solving (1) and (2), we get the point as

$$\left(\frac{c'-c}{m-m,}, \frac{mc'-m'c}{m-m'}\right).$$

Example 33:

What loci are represented by the equation

$$(x^2 - a^2)^2 (x^2 - b^2)^2 + c^4 (y^2 - a^2)^2 = 0.$$

Solution:

The given equation is

$$(x^2 - a^2)^2 (x^2 - b^2)^2 + c^4 (y^2 - a^2)^2 = 0.$$

This being the sum of two perfect squares, each term must be zero. Hence, we get

$$(x^2 - a^2) (x^2 - b^2)^2 (x^2 - b^2)^2 + c^4 (y^2 - a^2) = 0.$$

$$\Rightarrow \quad (x - a) (x + a) (x - b) (x + b) = 0 \qquad ...(1)$$

and $c^4 (y^2 - a^2)^2 = 0$

$$\Rightarrow \quad c^2 (y^2 - a^2) = 0$$

or $c^2 (y + a) (y - a) = 0.$ ...(2)

Equation no. (1) holds good for $x = \pm a$ or $x = \pm b$.

Equation no. (2) is satisfied by $y = \pm a$.

As both of these should be simultaneously satisfied, the given line represents 8 points which we get as a result of different combinations of (1) and (2), namely $(\pm a, \pm a)$, $(\pm b, \pm a)$.

Example 34:

Show also that the pair $ax^2 + 2hxy + by^2 + \lambda (x^2 + y^2) = 0$ is equally inclined to the same pair.

Solution:

The given equation is

$$ax^2 + 2hxy + by^2 + \lambda (x^2 + y^2) = 0 \qquad ...(1)$$

$$(a + \lambda) x^2 + 2hxy + (b + \lambda) y^2 = 0.$$

The equation of bisector will be

$$\frac{x^2 - y^2}{(a+\lambda)-(b+\lambda)} = \frac{xy}{h} \text{ or } \frac{x^2 - y^2}{a-b} = \frac{xy}{h}$$

This equation is independent of λ. Hence (2) has the same bisector for all values of λ.

Example 35:

Show that the equation $\cos 3\alpha\,(x^3 - 3xy^2) + \sin 3\alpha\,(y^3 - 3x^2y) + 3a\,(x^2 + y^2) - 4a^3 = 0$ *represents three straight lines forming an equilateral triangle. Prove also that its area is* $3\sqrt{3a^2}$.

Solution:

The given equation is

$$\cos 3\alpha\,(x^3 - 3xy^2) + \sin 3\alpha\,(y^3 - 3x^2) + 3a\,(x^2 + y^2) - 4a^3 = 0$$

Changing in polar co-ordinates by putting $x = r\cos\theta$, $y = r\sin\theta$, we get

$$\cos 3\alpha\,(r^3\cos^3\theta - 3r^3\cos\theta\sin^2\theta) + \sin 3\alpha$$
$$(r^3\sin^3\theta - 3\cos^2\theta\sin\theta) + 3a\,(r^2\cos^2\theta + r^2\sin^2\theta) - 4a^3 = 0$$

$$\Rightarrow r^3\cos 3\alpha\cos\theta\,(\cos^2\theta - 3\sin^2\theta)$$
$$+ r^3\sin 3\alpha\sin\theta\,(\sin^2 - 3\cos^2\theta) + 3ar^2 - 4a^3 = 0$$

As $\quad \cos^2\theta - \sin^2\theta = 4\cos^2\theta - 3$

and $\quad \sin^2\theta - 3\cos^2\theta = -(3 - 4\sin^2\theta)$,

we get $\quad r^3\cos 3\alpha.[4\cos^3\theta - 3\cos\theta]$

$$-r^2\sin 3\alpha\,[3\sin\theta - 4\sin^3\theta] + 3ar^2 - 4a^3 = 0$$

$$\Rightarrow \quad r^3\,[\cos 3\alpha\cos 3\theta - \sin 3\alpha\sin 3\theta] + 3ar^2 - 4a^3 = 0$$

$$\Rightarrow \quad r^3\cos(3\alpha + 3\theta) + 3ar^2 - 4a^3 = 0$$

$$\Rightarrow \quad \cos(3\alpha + 3\theta) = 4\left(\frac{a}{r}\right)^3 - 3\left(\frac{a}{r}\right).$$

As R.H.S. is of the form $4\cos^3\lambda - 3\cos l)$, the above equation may have 3 different solutions between 0 and 360° as

$$\frac{a}{r} = \cos(a + q) \qquad ...(1)$$

$$\frac{a}{r} = \cos\left(\alpha + \theta + \frac{2\pi}{3}\right) \qquad ...(2)$$

and $$\frac{a}{r} = \cos\left(\alpha + \theta + \frac{4\pi}{3}\right). \qquad ...(3)$$

Hence, the given line represent 3 straight lines whose equations are given by (1), (2) and (3). The acute angle between (1, 2), (2, 3) and (3, 1) is clearly supplementary of 2p/3 or 60°.

Hence, if they meet at A, B and C respectively, all the angles D,A,B,C are 60° each. Therefore, ABC is an equilateral triangle.

Again (1) is $\frac{a}{r} = \cos(\alpha + \theta)$.

$\Rightarrow \quad a = r(\cos\alpha \cos\theta - \sin\alpha \sin\theta)$

$\Rightarrow \quad a = r\cos\theta\cos\alpha - r\sin\theta\sin\alpha$

$\Rightarrow \quad x\cos\alpha - y\sin\alpha = a$ ...(4)

[as $x = r\cos\theta$ and $y = r\sin\theta$].

Similarly, (2) and (3) on changing into Cartesian co-ordinates become

$$x\cos\left(\alpha + \frac{2\pi}{3}\right) - y\sin\left(\alpha + \frac{2\pi}{3}\right) = a \quad ...(5)$$

and
$$x\cos\left(\alpha + \frac{4\pi}{3}\right) - y\sin\left(\alpha + \frac{4\pi}{3}\right) = a \quad ...(6)$$

respectively.

To get the point of intersection of (2) and (3); i.e., B, simplifying (5), we get

$$x\cos\alpha\cos\frac{2\pi}{3} - x\sin\alpha\sin\frac{2\pi}{3} - y\sin\alpha\cos\frac{2\pi}{3} - y\cos\alpha\sin\frac{2\pi}{3} = a$$

$$\Rightarrow \quad -x\cos\alpha.\frac{1}{2} - x\sin\alpha\frac{\sqrt{3}}{2} + y\sin\alpha\frac{1}{2} - y\cos\alpha\frac{\sqrt{3}}{2} = a$$

$$\Rightarrow \quad -\cos\alpha - x\sqrt{3}\sin\alpha + y\sin\alpha - y\sqrt{3}\cos\alpha = 2a. \quad ..(7)$$

Similarly, simplifying (6), we get

$$-x\cos\alpha + x\sqrt{3}\sin\alpha + y\sin\alpha - y\sqrt{3}\cos\alpha = 2a. \quad ...(8)$$

Adding (7) and (8), we get

$$-2x\cos\alpha + 2y\sin\alpha = 4a \text{ or } -x\cos\alpha + y\sin\alpha = 2a. \quad ..(9)$$

Subtracting (8) from (7), we get

$$-2x\sqrt{3}\sin\alpha - 2y\sqrt{3}\cos\alpha = 0$$

$$\Rightarrow \quad x\sin a + y\cos a = 0. \quad ...(10)$$

Solving (9) and (10), we get $x = -2a\cos\alpha$, $y = 2a\sin\alpha$.

Hence the co-ordinates of B are $(-2a\cos\alpha, 2a\sin\alpha)$.

Equation of AC is given by (4).

Perpendicular distance of B from AC

$$= \frac{-2a\cos^2\alpha - 2a\sin^2\alpha - a}{\sqrt{(\cos^2\alpha + \sin^2\alpha)}} = 3a \text{ (omitting – ve sign).}$$

As the triangle ABC is equilateral one, Its area = $\frac{(3a)^2}{\sqrt{3}} = 3\sqrt{3a}^{2}$.

[$\therefore$ if p is the height of an equation triangle, its area will be $p^2/\sqrt{3}$].

Example 36:

Prove that two of the lines represented by the equation $ax^4 + bx^3y + cx^2y^2 + dxy^3 + ay^4 = 0$ will bisect the angles between the other two if $c = 6a = 0$ and $b + d = 0$.

Solution:

The equation is given as

$$ax^4 + bx^3y + cx^2y^2 + dxy^3 + axy^3 + ay^4 = 0. \qquad ...(1)$$

The equation being homogeneous equation of fourth degree will represent 4 straight lines. Let one pair of these be

$$(px^2 + 2qxy + ry^2) \qquad ...(2)$$

The equations of the bisectors of the angles between the lines given by (2) will be $\frac{x^2 - y^2}{p - r} = \frac{xy}{q}$.

$$\theta\,(x^2 - y^2) - (p - r)\,xy = 0.$$

As (1) represents by hypothesis 2 pairs of straight lines, one pair of which is the bisector of the angles between the other set, we must have

$$ax^4 + bx^3y + cx^2y^2 + dxy^3 + ay^4 \equiv (px^2 + 2qxy + ry^2)\,[qx^2 - qy^2 - (p - r)\,xy].$$

Comparing the coefficient of different terms in order, we get

$$\frac{a}{pq} = \frac{b}{-p^2 + pr + 2q^2} = \frac{c}{-3q\,(r - p)} = \frac{d}{-2p^2 - pr + r^2} = -\frac{a}{qr}.$$

By 1 and 5 ratio, r = –p.

Putting in 2 and 4 ratios.

$$-\frac{b}{p^2 - p^2 = 2q^2} = \frac{d}{2q^2 + p^2 + p^2}$$

or $\qquad b = -d$ or $b + d = 0$.

Again by 1 and 3 ratios

$$\frac{a}{pq} = \frac{c}{3p.(-p-p)} \text{ or } a = -\frac{c}{6} \text{ or } 6a + c = 0.$$

Example 37:

If the equation $ax^2 + 2hxy + by^2 + 2gx + 2fy + c = 0$, represent a pair of straight lines, prove that the equation to the third pair of straight lines passing through the points where these meet the axes is

$$ax^2 - 2hxy + by^2 + 2gx + 2fy + c + \frac{4fg}{c}xy = 0$$

Solution:

We are given the equations as

$$ax^2 + 2hxy + by^2 + 2gx + 2fy + c = 0. \qquad ...(1)$$

The equation of y-axis is $x = 0$ and x-axis is $y = 0$. Hence the combined equation will be $xy = 0$. ...(2)

Equation of the curve passing through the point of intersection of (1) and (2) will be

$$ax^2 + 2hxy + by^2 + 2gx + 2fy + 2fy + 2\lambda xy = 0$$

$$\Rightarrow \quad ax^2 + 2(h + \lambda)\, xy + by^2 + 2gx + 2fy + c\ 0. \qquad ...(3)$$

If (3) represent two straight lines, then its discriminant must be zero.

So $abc + 2.f.g.\ (h + \lambda) - af^2 - bg^2 - c\,(h + \lambda)^2 = 0$

$$\Rightarrow \quad abc + 2fgh - af^2 - bg^2 - ch^2 + 2\lambda\,(fg - ch) - c\lambda^2 = 0 \quad ...(4)$$

As (1) represents 2 straight lines

$$abc + 2fgh - af^2 - bg^2 - ch^2 = 0.$$

Putting in (4), we get $2\lambda\,(fg - ch) - cl = 0$ or $\lambda = \frac{2\,(fg - ch)}{c}$.

Putting in (3), we get

$$ax^2 + 2\left[h + \frac{2\,(fg - ch)}{c}\right] xy + by^2 + 2gx + 2fy + c = 0$$

$$\Rightarrow \quad ax^2 + \frac{4fg - 2ch}{c}\, xy + by^2 + 2gx + 2fy + c = 0$$

$$ax^2 - 2\,hxy + by^2 + 2gx + 2fy + c + \frac{4fg}{c}xy = 0.$$

Example 38:

Prove that the straight lines $ax^2 + 2hxy + by^2 = 0$, make equal angles with the axis of x if $h = a \cos \omega$, the axes being inclined at an angle ω.

Solution:

Let the equations represented by $ax^2 + 2hxy + by^2 = 0$ by

$$y - m_1x = 0$$

and $y - m_2x = 0$,

$$m_1 + m_2 = -2h/b$$

and $m_1m_1 = a/b$.

If the lines make equal angles with x-axis, one angle being θ, other must be (180 – θ).

So $$\tan \theta = \frac{m_1 \sin \omega}{1 + m_1 \cos \omega}, \qquad ...(1)$$

and $$\tan (180 - \theta) = \frac{m_2 \sin \omega}{1 + m_2 \cos \omega}. \qquad ...(2)$$

By (1) and (2), we get $$\frac{m_1 \sin \omega}{1 + m_1 \cos \omega} = -\frac{m_2 \sin \omega}{1 + m_2 \cos \omega}.$$

[as tan (180 – θ) = – tan θ]

$$\Rightarrow \quad m_1 + m_2 + 2m_1m_2 \cos \omega = 0. \qquad ...(3)$$

Putting the values of $m_1 + m_2$ and m_1m_2, we get

$$\frac{-2h}{b} + \frac{2a \cos w}{b} = 0$$

or $$h = a \cos \omega.$$

Example 39:

If the axes be inclined at an angle ω, show that the equation $x^2 + 2xy \cos \omega + y^2 \cos 2\omega = 0$, represents a pair of perpendicular straight lines.

Solution:

Let the given equation be

$$x^2 + 2xy \cos \omega + y^2 \cos 2\omega = 0 \qquad ...(1)$$

represent 2 straight lines $y = m_1x$ and $y = m_2x$. Then the combined equation will be

$$(y - m_1x)(y - m_2x) = 0. \qquad ...(2)$$

Comparing the coefficients of different terms in (1) and (2)

$$\frac{1}{m_1 m_2} = \frac{2\cos\omega}{-(m_1 + m_2)} = \frac{\cos 2\omega}{1}\cdot$$

$$\therefore \qquad m_1 m_2 = \frac{1}{\cos 2\omega} \text{ and } m_1 + m_2 = \frac{2\cos\omega}{\cos 2\omega}$$

If lines given by (2) are perpendicular, then

$1 + (m_1 + m_2)\cos\omega + m_1 m_2 = 0.$

Putting the values of $m_1 + m_2$ and $m_1 m_2$ in L.H.S. we have

$$1 - \frac{2\cos\omega}{\cos 2\omega}\cos\omega + \frac{1}{\cos 2\omega}$$

$$= \frac{1}{\cos 2\omega}\left(\cos 2\omega - 2\cos^2\omega + 1\right) = 0 \quad (\because \cos 2\omega = 2\cos^2\omega - 1).$$

Hence, the lines represented by (1) are perpendicular to each other.

Example 40:

If the equation $ax^2 + 2hxy + by^2 + 2gx + 2fy + c = 0$, represent two straight lines, prove that the square of the distance of their point of intersection from the origin is $\dfrac{c(a+b) - f^2 - g^2}{ab - h^2}$.

Solution:

The given equation is

$$ax^2 + 2hxy + by^2 + 2gx + 2fy + c = 0 \qquad ...(1)$$

be $\qquad y - m_1x - c_1 = 0 \qquad ...(2)$

and $y - m_2x - c_2 = 0 \qquad ...(3)$

The combined must be similar to (1), so

$ax^2 + 2hxy + by^2 + 2gx + 2fy + c \Rightarrow b(y - m_1x - c_1)(y - m_2x - c_2)$

(we multiply R.H.S. by b so as to make the coefficient of term say y^2 on R.H.S. equal to coefficient of y^2 on L.H.S. So the coefficients must also be equal). So equating other coefficients, we get

$$m_1 + m_2 = -\frac{2h}{b} \qquad ...(4)$$

$$m_1 + m_2 = \frac{a}{b} \qquad ...(5)$$

$$c_1 m_2 + c_2 m_1 = \frac{2g}{b} \qquad ...(6)$$

$$c_1 + c_2 = -\frac{2f}{b} \qquad \text{...(7)}$$

and $$c_1 c_2 = \frac{c}{b}. \qquad \text{...(8)}$$

The point of intersection of (2) and (3) will be given by

$$\frac{y}{m_1 c_2 - m_2 c_1} = \frac{x}{-c_1 + c_2} = \frac{1}{-m_2 + m_1}$$

Hence the co-ordinates of the point of intersection are

$$\left(\frac{c_2 - c_1}{m_1 - m_2}, \frac{m_1 c_2 - m_2 c_1}{m_1 - m_2}\right).$$

The square of distance of this point from the origin (0, 0) is

$$\left[\frac{c_2 - c_1}{m_1 - m_2}\right]^2 + \left[\frac{m_1 c_2 - m_2 c_1}{m_1 - m_2}\right]^2$$

$$= \frac{\left[(c_1 + c_2)^2 - 4c_1 c_2\right] + (m_1 c_2 + m_2 c_1)^2 - 4m_1 m_2 c_1 c_2}{(m_1 + m_2)^2 - 4m_1 m_2}$$

$$= \frac{\left[\frac{4f^2}{b^2} - \frac{4c}{b}\right] + \left[\frac{4g^2}{b^2} - \frac{4ac}{b^2}\right]}{\frac{4h^2}{b^2} - \frac{4a}{b}}$$

$$= \frac{f^2 - bc + g^2 - ac}{h^2 - ab} = \frac{f^2 + g^2 - c(b + a)}{h^2 - ab} = \frac{c(a + b) - f^2 - g^2}{ab - h^2}.$$

Example 41:

Show that two of the straight lines represented by the equation $ay^4 + bxy^3 + cx^2y^2 + dx^3y + ex^4 = 0$ will be at right angles if

$$(b + d)(ad + be) + (e - a)^2 (a + c + e) = 0.$$

Solution:

The given equation is $ay^4 + bxy^3 + cx^2y^2 - dxy^3 + ex^4 = 0$.

If two of the lines represented by (1) at right angles, their equation must be of the form $x^2 + pxy - y^2 = 0$ as the sum of the coefficients of x^2 and y^2 is zero. Hence let

$$ay^4 + bxy^3 + cx^2y^2 + dyx^3 + ex^4 \equiv (x^2 + pxy - y^2)(ex^2 + qxy - ay^2)$$

As the coefficient of x^4 on both sides is same, other coefficients with also be equal.

Now compare different coefficients.

Comparing xy^3, we have $-pa - q = b$. ...(2)

Comparing x^2y^2, we have $-a + pq - e = c$. ...(3)

Comparing x^3y, we have $q + eq = d$

Now eliminate p and q from (2), (3) and (4).

From (2) and (4) $p = \frac{b+d}{e-a}$

and $q = -\frac{da+eb}{e-a}$.

Substituting these values in (3), we get

$$-a - \frac{b+d}{e-a} \cdot \frac{da+eb}{e-a} - e = c$$

$$\Rightarrow \quad (a + e + c)(e - a)^3 + (b + d)(ad + be) = 0.$$

Example 42:

Hence find the locus of the orthocentre of a triangle of which two sides are given in position and whose third side goes through a fixed point.

Solution:

Let the line $\lambda x + my = 1$ passes through some fixed point $P \equiv (\alpha, \beta)$. Then $l\alpha + m\beta = 1$. ...(1)

If (x', y') be the orthocentre,

$$\frac{x'}{l} = \frac{y'}{m} = \frac{a+b}{am^2 - 2hl\,m + bl^2} + \frac{x'+y'}{l+m} \qquad ...(2)$$

As $\frac{x'}{l} = \frac{y'}{m} = \frac{\alpha x'}{\alpha l} = \frac{\beta y'}{\beta m} = \frac{\alpha x'+\beta y'}{\alpha l+\beta m} = \alpha x'+\beta y'$ [as $al + \beta m = 1$ by (1)]

$$\therefore \quad \frac{x'}{l} = \frac{y'}{m} = \alpha x'+\beta y'.$$

$$\therefore \quad l = \frac{x'}{\alpha x'+\beta y'}, \; m = \frac{y'}{\alpha x'+\beta y'}.$$

Putting these values in (2), we get

$$\frac{a+b}{a\left(\frac{y'}{\alpha x'+\beta y'}\right)^2-2h\frac{x'}{\alpha x'+\beta y'}\cdot\frac{y'}{\alpha x'+\beta y'}+b\left(\frac{x'}{\alpha x'+\beta y'}\right)^2}$$

$$=\frac{x'+y'}{\frac{x'}{\alpha x'+\beta y'}+\frac{y'}{\alpha x'+\beta y'}}$$

$$\Rightarrow \quad \frac{(a+b)(\alpha x'+\beta y')^2}{ay'^2-2hx'y'+bx'^2}=\frac{(x'+y')(\alpha x'+\beta y')}{(x'+y')}.$$

Simplifying and generalising for x' and y', we get the required locus as $(a + b)(\alpha x + \beta y) = ay^2 - 2hxy + bx^2$.

Example 43:

Prove that one of the lines represented by the equation $ax^3 + bx^2y + cxy^2 + dy^3 = 0$, will bisect the angle between the other two if $(3a + c)^2 (bc + 2cd - 3ad) = (b + 3d)^2 (bc + 2ab - 3ad)$.

Solution:

The given line is $ax^3 + bx^2y + cxy^2 + dy^3 = 0$ ...(1)

This being homogeneous cubic equation represents three straight lines each passing through the origin.

Let these lines be $y - m_1x = 0$, ...(2)

$y - m_2x = 0$, ...(3)

and $y - m_3x = 0$. ...(4)

Let the equation (4) be the bisector of the angle between (2) and (3). Then, if the angles that the lines (2), (3) and (4) make with x-axis be θ_1, θ_2 and θ. clearly $2\theta = \theta_1 + \theta_2$ or $\tan 2\theta = \tan(\theta_1 + \theta_2)$

$$\Rightarrow \quad \frac{2\tan\theta}{1-\tan^2\theta}=\frac{\tan\theta_1+\tan\theta_2}{1-\tan\theta_1\tan\theta_2}=\frac{3m_3}{1-m_3^2}=\frac{m_1+m_2}{1-m_1m_2}. \quad ...(5)$$

Again equation no. (1) is same as represented by (2), (3) and (4).

Hence $ax^3 + bx^2y + cxy^2 + dy^3 \equiv (y - m_2x)(y - m_3x)$.

Comparing the coefficients, we get

$$\frac{1}{d}=\frac{-(m_1+m_2+m_3)}{c}=\frac{m_1m_2+m_2m_3+m_3m_1}{b}=\frac{-m_1m_2m_3}{a}$$

$$\Rightarrow \quad m_1 + m_2 + m_3 = -\frac{c}{d}, \quad ...(6)$$

$$m_1 m_2 + m_2 m_3 + m_3 m_1 = \frac{b}{d} \qquad ...(7)$$

and $$m_1 m_2 m_3 = -\frac{a}{d} \qquad ...(8)$$

Putting the value of $m_1 + m_2$ from (6) in (5), we get

$$\frac{2m_3}{1-m_3^2} = \frac{-\frac{c}{d}-m_3}{1-m_1 m_2} = \frac{-c-dm_3}{d-dm_1 m_2}$$

$$\Rightarrow \quad 2m_3d - 2dm_1m_2m_3 = -c - dm_3 = cm_3^2 + dm_3^2$$

$$\Rightarrow \quad dm_3^2 + cm_3^2 - c + 2d\left(-\frac{a}{d}\right) = 0 \qquad ...(9)$$

$$\left\{\because \ m_1 m_2 m_3 = -\frac{a}{d} \text{ by (8)}\right\}$$

Now dividing (1) by x^2, we get, $a + b\frac{y}{x} + c\left(\frac{y}{x}\right)^2 + d\left(\frac{y}{x}\right)^3 = 0.$

As $y = m_3$ and (10), on substraction, we get m_3 (b + 3d) = – (3a + c)

So $$a + bm_3 + cm_3^2 + dm_3^2 = 0. \qquad ...(10)$$

From (9) and (10), on substraction, we get m_3 (b + 3d) = – (3a + c)

or $$m_3 = \frac{3a+c}{b+3d}.$$

Substituting this value in (10) and simplifying, we get the required result as $(3a + c)^2 (bc + 2cd - 3ad) = (b + 3d)^2 (bc + 2ab - 3ad)$.

Example 44:

Show that the orthocentre of the triangle formed by the straight lines $ax^2 + 2hxy + by^2 = 0$ and $lx + my + l$ is a point (x', y') such that

$$\frac{x'}{l} = \frac{y'}{m} = \frac{a+b}{am^2 - 2hlm + bl^2}.$$

Solution:

Let the equation $ax^2 + 2hxy + by^2 = 0$ represent OB and OA respectively as

$$y = m_1x \qquad ...(1)$$

and $$y = m_2x \qquad ...(2)$$

Then $m_1 + m_2 = -\frac{2h}{b}$ and $m_1 m_2 = \frac{a}{b}$

The line AB is given as $lx + my = 1$. ...(3)

Solving (1) and (3), we get $x = \frac{1}{l+mm_1}$ $y = \frac{m_1}{l+mm_1}$

Hence the co-ordinates of B are $\left(\frac{1}{l+mm_1} \frac{m_1}{l+mm_1}\right)$.

The equation of the line perpendicular to OB i.e., (2), passing to OB i.e., (2), passing through opposite vertex A, will be

$$y - \frac{1}{l+mm_1} = -\frac{1}{m_2}\, x - \left(\frac{1}{l+mm_1}\right)$$

or $(l + mm_1)\, x\, m_2\, (l + mm_1)\, y - (m_1 m_2 + 1) = 0$.

As the line passes through the orthocentre (x', y'), so it will satisfy it. Hence,

$$(l + mm_1)\, x' + m_2\, (l + mm_1)\, y' - (m_1 m_2 + 1) = 0. \quad ...(4)$$

Again, the line perpendicular to AB passing through the vertex O (0, 0) will be $mx - l\,y = 0$.

As it also passes through the orthocentre (x', y'), hence

$$mx' - ly' = 0. \quad ...(5)$$

To solve (4) and (5), putting value of y' from (5) in (4); we get

$$\left(l+mm_1\right) x' + m_2 \left(l+mm_1\right) \frac{mx'}{l} - \left(m_1\, m_2 + 1\right) = 0.$$

$$\Rightarrow \quad \frac{x'}{l}\left[l^2 + l\,mm_1\, l + mm_2\, l + m^2\, m_1\, m_2\right] = 1 + m_1\, m_2$$

$$\Rightarrow \quad \frac{x'}{l}\left[l^2 + l\,m\left(m_1 + m_2\right) + m_1\, m_2\right] = 1 + m_1\, m_2.$$

Putting the values of $m_1 + m_2$ and $m_1 m_2$ as $-\frac{2h}{b}$ and $\frac{a}{b}$ respectively, we get

$$= \frac{x'}{l}\left[l^2 - l\,m\frac{2h}{b} + m^2\frac{a}{b}\right] = 1 + \frac{a}{b} \text{ or } \frac{x'}{l} = \frac{a+b}{l^2 b - 2hlm + am^2}$$

Similarly putting the value of x' from (5) in (4), we get

$$\frac{y'}{l} = \frac{a+b}{l^2 b - 2hlm + am^2}.$$

Hence
$$\frac{x'}{l} - \frac{y'}{m} - \frac{a+b}{l^2 b - 2hlm + am^2}.$$

Example 45:

Transform to axes inclined at 45° to the original axes the equations (1) $x^2 - y^2 = a^2$, (2) $17x^2 - 16xy + 17y^2 = 225$, and (3) $y^4 + x^4 + 6x^2 y^2 = 2$.

Solution:

(i) Given equation is $x^2 = y^2 = a^2$. The axes are turned to an angle of 45°, hence putting (x' cos 45° – y' sin 45°) for x and

(x' sin 45° + y' cos 45°) for y, we get

$$\Rightarrow \quad \frac{1}{2}[(x' - y')^2 - (x' + y')^2] = a^2 \therefore \cos 45° = \sin 45° = \frac{1}{\sqrt{2}}$$

$$\Rightarrow \quad 2x'y' + a^2 = 0.$$

(ii) If the curve is $17x^2 - 16xy + 17y^2 = 225$, by the same substitution, we get

$$17(x' \cos 45 - y' \sin 45)^2 - 16(x' \cos 45 - y' \sin 45)$$
$$\times (x' \sin 45 + y' \cos 45) + 17(x' \sin 45 + y' \cos 45)^2 = 225$$

$$\Rightarrow \quad \frac{1}{2}\left[18x^2 + 50y^2\right] = 225 \text{ or } 9x^2 + 25y^2 = 225.$$

(iii) Curve is $y^4 + x^4 + 6x^2y^2 = 2$.

By the same substitution, we get

$$(x' \cos 45 + y' \sin 45)^4 + (x' \sin 45 - y' \cos 45)^4$$
$$+ 6(x' \cos 45 - y' \sin 45)^2 (x' \sin 45 + y' \cos 45)^2 = 2$$

$$\therefore \quad x'^4 + y'^4 = 1 \text{ (on simplification).}$$

Example 46:

Transform to axes inclined at an angle α to the original axes the equations (1) $x^2 + y^2 = r^2$, and (2) $x^2 + 2xy \tan 2\alpha - y^2 = a^2$.

Solution:

The given equation is $x^2 + y^2 = r^2$.

(i) The axes are turned to an angle α. Substituting the values of x and y in x', and α, we get, $(x' \cos \alpha - y' \sin \alpha)^2 + (x \sin + y' \cos \alpha)^2 = r^2$.

Simplifying, we get $x'^2 + y'^2 = r^2$.

(ii) $x^2 + 2xy \tan 2\alpha - y^2 = a^2$. By the same substitution, we get

$$(x' \sin \alpha - y' \sin \alpha)^2 + 2(x' \cos \alpha - y' \sin \alpha)$$
$$\times (x' \sin \alpha + y' \cos \alpha) \tan 2\alpha - (x' \sin \alpha + y' \cos \alpha)^2 = a^2$$

$$\Rightarrow \quad x'^2 (\cos^2 \alpha - \sin^2 \alpha) - y'^2 (\cos^2 \alpha - \sin^2 \alpha) - 2x'y' \sin \alpha \cos \alpha$$
$$+ [2 \sin \alpha \cos \alpha (x'^2 - y'^2) + x'y' (\cos^2 \alpha - \sin^2 \alpha] \tan 2\alpha = a^2$$

$$\Rightarrow \quad x'^2 \cos 2\alpha - y'^2 \cos 2\alpha - x'y' \sin 2\alpha$$
$$= \left[\left(x'^2 - y'^2\right) \sin 2\alpha + x'y' \cos 2\alpha\right] \frac{\sin 2\alpha}{\cos 2\alpha} = a^2$$

$$\Rightarrow \quad (x'^2 - y'^2) \cos^2 2\alpha - x'y' \sin 2\alpha + (x'^2 - y'^2) \sin^2 2\alpha$$
$$+ x'y' \sin 2\alpha = \alpha^2 \cos 2\alpha.$$

$$\Rightarrow \quad x'^2 - y'^2 = a^2 \cos 2\alpha.$$

Example 47:

Transform the equation $y^2 + 4y \cot \alpha - 4x = 0$ from rectangular axes to oblique axes meeting at an angle α, the axis of x being kept the same.

Solution:

The given equation is $y + 4y \cot \alpha - 4x = 0$. ...(1)

By hypothesis, $\omega = 90°$, $\omega' = \alpha$, $\theta = 0$.

For transformation $x \sin 90° = x' \sin (90 - 0) + y' \sin (90 - \alpha - 0)$

$\Rightarrow \quad x = x' + y' \cos \alpha$ ...(2)

and $\quad y \sin 90° = x' \sin 0 + y' \sin (\alpha + 0)$

$\Rightarrow \quad y = y' \sin \alpha$. ...(3)

Putting the values from (2) and (3) in (1), we get

$$(y' \sin \alpha)^2 + 4y' \sin \alpha . \frac{\cos \alpha}{\sin \alpha} - 4 . (x' + y' \cos \alpha) = 0$$

$\Rightarrow \quad y'^2 \sin^2 \alpha + 4y' \cos \alpha - 4x' - 4y' \cos \alpha = 0$

$\Rightarrow \quad y'^2 \sin^2 \alpha + 4x'$

or $\quad y'^2 = 4x' \operatorname{cosec}^2 \alpha$.

Example 48:

Transform the equation $2x^2 + 3\sqrt{3xy} + 3y^2 = 2$ from axes inclined at $30°$ to rectangular axes, the axis of remaining unchanged.

Solution:

The given equation is $2x^2 + 3\sqrt{3xy} + 3y^2 = 0.$ ...(1)

By hypothesis, ω = 30' = 90°, θ = 0, hence we have

$$x \sin 30 = \omega' = 90°,$$

$$\theta = 0, \text{ hence we have}$$

$\Rightarrow$ $$\frac{1}{2}.x = x'.\frac{1}{2} - \frac{\sqrt{3}}{2}.t' \text{ or } x = x' - \sqrt{3y'}$$ ...(2)

and $y \sin 30 = x' \sin = 0 + y' \sin (90 + 0)$

$\Rightarrow$ Substituting from (2) and (3) in (1), we get

$$2\,(x' - \sqrt{3y'}\,)^2 + 3\sqrt{3}\,(x' - \sqrt{3y}\,') + 3\,(2y)^2 = 2$$

$\Rightarrow$ $x'^2 + \sqrt{3.x'y'} = 1$ (on simplification).

Example 49:

Transform the equation $x^2 + xy + y^2 = 8$ from axes inclined at 60° to axes bisecting the angles between the original axes.

Solution:

The origin axes are inclined at angle of 60°.

Bisectors of the angles will make angles of 30° and 120° from original axis of x. Hence here ω = 30°, ω' = 90°, θ = 30°.

For transformation,

$$x \sin 60° = x' \sin (60° - 30°) + y' \sin (60° - 90° - 30°)$$

$\Rightarrow$ $$x.\frac{\sqrt{3}}{2} = x'.\frac{1}{2} - y'\frac{\sqrt{3}}{2} \text{ or } x = \frac{x'}{\sqrt{3}} - y'.$$ ...(1)

And $y \sin 60° = x' \sin 30 + y' \sin (90 + 30)$

$\Rightarrow$ $$y.\frac{\sqrt{3}}{2} = x'.\frac{1}{2} + y'\frac{\sqrt{3}}{2} \text{ or } y = \frac{x'}{\sqrt{3}}\,x' + y'$$ ...(2)

The original equation is given as $x^2 + xy + y^2 = 8$.

Substituting the values of x and y from (1) and (2), we get

$$\left(\frac{x'}{\sqrt{3}} - y'\right)^2 + \left(\frac{x'}{\sqrt{3}} - y'\right)\left(\frac{x'}{\sqrt{3}} + y'\right) + \left(\frac{x'}{\sqrt{3}} + y'\right)^2 = 8$$

$\Rightarrow$ $x'^2 + y'^2 = 8$ (on simplification).

Example 50:

The equation to a straight line referred to axes inclined at 30° to one another is y = 2x + 1. Find its equation referred to axes inclined at 45°, the origin and axis of x being unchanged.

Solution:

The given equation is y = 2x + 1. ...(1)

by hypothesis, w = 30°, w' = 45° and q = 0. So we have the substitution x sin 30° = x' sin (30 – 0) + y' sin (30 – 45 – 0)

$$\Rightarrow \quad \frac{1}{2}x = x'\frac{1}{2} - y'\frac{\sqrt{3}-1}{2\sqrt{2}} \text{ or } x = x' - \frac{\sqrt{3}-1}{\sqrt{2}}y' \quad ...(2)$$

and y sin 30° = x' sin 0 + y' sin 45

$$\Rightarrow \quad y.\frac{1}{2} = y'.\frac{1}{\sqrt{2}} \text{ or } y = \sqrt{2}.y' \quad ...(3)$$

Putting the values from (2) and (3) in (1), we get

$$\sqrt{2}.y' = 2.\left(x' - \frac{\sqrt{3}-1}{\sqrt{2}}y'\right) + 1$$

$$2x' - \sqrt{6y'} + 1 = 0.$$

Example 51:

If x and y be the coordinates of a point referred to a system of oblique axes, and x' and y' be its coordinates referred to another system of oblique axes with the same origin, and if the formulae of transformation be

$$x = mx' + ny'$$

and $\quad y = m'x' + n'y'$,

prove that $\quad \dfrac{m'^2 + m'^2 - 1}{n'^2 + n'^2 - 1} = \dfrac{mm'}{nn'}$.

Solution:

Let the original axes be inclined at an angle of ω, the origin being O. If there is any point P whose co-ordinates with respect to these axes be (x, y); then

$$OP^2 = x^2 + y^2 + 2xy \cos \omega.$$

...(1)

If the axes are changed and the angle between the new axes be ω', origin remaining the same, let the co-ordinates of P with respect to new axes be (x', y')' then

$$OP^2 = x'^2 + y'^2 + 2x'y, \cos \omega'. \quad ...(2)$$

By (1) and (2), we get

$$x^2 + y^2 + 2xy \cos \omega = x'^2 + y'^2 + 2x'y' \cos \omega'. \quad ..(3)$$

As the transformation are given as,

$$x = mx' + ny' \text{ and } y = m'x' + n'y'.$$

Putting these values in (3), we get

$$(mx' + ny)^2 + (m'x' + n'y')^2 + 2 (mx' + ny') (m'x' + n'y') \cos \omega.$$
$$= x'^2 + y'^2 + 2x'y' \cos \omega. \quad ...(4)$$

Comparing the coefficients of x'^2 on both sides of (4), we have

$$m^2 + m'^2 = 2mm' \cos \omega = 1.$$

$$\Rightarrow \quad m^2 + m'^2 - 1 = -2nn' \cos \omega \quad ...(6)$$

Dividing (6) by (5), we get

Comparing the coefficients of y'^2 in (4), we have

$$n^2 + n'^2 = 2mm' \cos \omega = 1.$$

$$\Rightarrow \quad n^2 + n'^2 - 1 = -2nn' \cos \omega \quad ...(6)$$

Dividing (6) by (5), we get

$$\frac{n^2 + n'^2 - 1}{m^2 + m'^2 - 1} = \frac{nn'}{mm'},$$

$$\Rightarrow \quad \frac{m^2 + m'^2 - 1}{mm'} = \frac{n^2 + n'^2 - 1}{nn'}.$$

Example 52:

Transform to parallel axes through the point (1, –2) the equations (1) $y^2 - 4x + 4y + 8 = 0$ and (2) $2x^2 + y^2 - 4x = 0$.

Solution:

(i) The equation is $y^2 - 4x + 4y + 8 = 0$. The origin is transferred to (1, –2). So the new equation will be

$$(y' - 2)^2 - 4 (x' + 1) + 4 (y' - 2) + 8 = 0$$

$$\Rightarrow \quad y'^2 - 4y' + 4 - 4x' - 4 + 4y' - 8 + 8 = 0$$

$$\Rightarrow \quad y'^2 = 4x'.$$

(ii) The equation is $2x^2 + y^2 - 4x + 4y = 0$.

Transferring the origin to (1, –2), we get

$$2 (x' +1)^2 + (y' - 2)^2 - 4 (x' + 1) + 4 (y' -2) = 0$$

$\Rightarrow \quad 2x'^2 + 4x' + 2 + y'^2 - 4y' + 4 - 4 + 4y' - 8 = 0$

$\Rightarrow \quad 2x'^2 + y'^2 = 6.$

Example 53:

What does the equation $(x - a)^2 + (y - b)^2 = c^2$, become when it is transferred to parallel axes through (1) the point $(a - c, b)$, (2) the point $(a, b, - c)$?

Solution:

The curve is given as $(x - a)^2 + (y - b)^2 = c^2$.

(i) Transferring the origin to the point $(a - c, b)$, we get

$$[x' + (a - c) - a]^2 + [y' + b - b)^2] = c^2$$

$\Rightarrow \quad x'^2 + c^2 - 2x'c + y'^2 = c^2$ or $x'^2 + y'^2 = 2cx'$.

(ii) Transferring the origin to the point $(a, b - c)$, we get

$$[x' + a - a]^2 + [y' + b - c - b]^2 = c^2$$

$\Rightarrow \quad x'^2 + y'^2 + c^2 - 2y'c = c^2$

or $\quad x'^2 + y'^2 = 2y'c.$

Example 54:

What does the equation $(a - b)(x^2 + y^2) - 2abx = 0$, become if the origin be moved to the point $(ab/a - b, 0)$?

Solution:

The equation is $(a - b)(x^2 + y^2) - 2abx = 0$.

The origin is transferred to $(ab/a - b, 0)$; hence the equation becomes

$$(a-b)\left[\left\{x'+\frac{ab}{a-b}\right\}^2 + \{y'+0\}^2\right] - 2ab.\left[x'+\frac{ab}{a-b}\right] = 0$$

$$\Rightarrow \quad (a-b)\left(x'^2+y'^2\right)+\frac{a^2\,b^2}{a-b} + lab.x' - 2abx' - \frac{2a^2\,b^2}{a-b} = 0$$

$$\Rightarrow \quad (a-b)\left(x'^2+y'^2\right)+\frac{a^2\,b^2}{a-b} \text{ or } (a-b)^2\left(x'^2+y'^2\right) = a^2\,b^2.$$

Example 55:

Find the angle through which the axes may be turned so that the equation $Ax + By + C = 0$, may be reduced to the form x = constant and determine the value of this constant.

Solution:

Let the required angle be α. Then turning the axes to the angle θ, origin being the same, the given equation $Ax + By + C = 0$ changes to $A(x' \cos \alpha - y' \sin \alpha) + B(x' \sin \alpha + y' \cos \alpha) + C = 0$.

or $\quad x'(A \cos \alpha + B \sin \alpha) - y'(A \sin \alpha - B \cos \alpha) + C = 0 \quad ...(1)$

If the equation takes the form x = constant, then the coefficient of y' must be zero; hence $A \sin \alpha - B \cos \alpha = 0$ or $A \sin \alpha = B \cos \alpha$ or $\tan \alpha = B/A$.

Hence the axes should be changed to an angle $\tan^{-1}(B/A)$.

Putting this value in (1), we get

$$x'\left[A.\frac{A}{\sqrt{(A^2+B^2)}}+B.\frac{B}{\sqrt{(A^2+B^2)}}\right]+C=0,$$

$\therefore$ if $\tan \alpha = \dfrac{B}{A}$ then $\cos \alpha = \dfrac{A}{\sqrt{(A^2+B^2)}}$ and $\sin \alpha = \dfrac{B}{\sqrt{(A^2+B^2)}}$

$$\Rightarrow \qquad x = -\frac{C\sqrt{(A^2+B^2)}}{A^2+B^2} = -\frac{C}{\sqrt{(A^2+B^2)}}$$

So required constant is $-\dfrac{C}{\sqrt{(A^2+B^2)}}$.

Example 56:

If the axes be turned through an angle $\tan^{-1} 2$, what does the equation $4xy - 3x^2 = a^2$ become ?

Solution:

The given equation is $4xy - 3x^2 = a^2$ and the angle $a = \tan^{-1} 2$ or $\tan \alpha = 2$; hence $\sin \alpha = 2/\sqrt{5}$ and $\cos \alpha = 1/\sqrt{5}$.

Substituting the appropriate values of x and y, we get

$$4(x' \cos \alpha - y' \sin \alpha)(x' \sin \alpha + y' \cos \alpha) - 3(x' \cos \alpha - y' \sin \alpha)^2 = a^2$$

$$\Rightarrow \qquad 4\left[\frac{1}{\sqrt{5}}x' - \frac{1}{\sqrt{5}}y'\right]\left[\frac{2}{\sqrt{5}}x' + \frac{1}{\sqrt{5}}y'\right] - 3\left[\frac{1}{\sqrt{5}}x' - \frac{2}{\sqrt{5}}y'\right]^2 = a^2$$

$\Rightarrow \qquad x'^2 - 4y'^2 = a^2$ (on simplification).

Example 57:

By transferring so parallel axes through a properly chosen point (h, k), prove that the equation $12x^2 - 10xy + 2y^2 + 11x - 5y + 2 = 0$, can be reduced to one containing only terms of the second degree.

Solution:

The given equation is

$$12x^2 - 10xy + 2y^2 + 11x - 5y + 2 = 0.$$

Let the origin be transferred to (h, k) axes being parallel to the previous axes; then the equation becomes

$$12(x' + h)^2 - 10(x' + h)(y' + k)^2 + 11(x' + h) - 5(y' + k) + 2 = 0$$

$$\Rightarrow 12x'^2 + 12h^2 + 24x'h - 10x'y' - 10x'k - 10y'h - 10hk + 2y'^2$$
$$+ 2k^2 + 4y'k + 11x' + 11h - 5y' - 5k + 2 = 0 \quad ...(1)$$

$$\Rightarrow 12x'^2 + 2y'^2 - 10x'y' + x'(24h - 10k + 11) + y'(-10h + 4k - 5)$$
$$+ 12h^2 - 10hk + 2k^2 + 11h - 5k + 2 = 0.$$

If this equation contains the terms of x^2 and y^2 and constant terms only, then the coefficients of x' and y' must be zero.

So $\quad 24h - 10k + 11 = 0 \quad ...(2)$

and $\quad (-10h + 5k - 5) = 0.$

Solving (2) and (3), we get $h = \frac{3}{2}$ and $k = -\frac{5}{2}$

Hence the required point is $\left(-\frac{3}{2}, -\frac{5}{2}\right)$,

If we substitute the values in (1), the equation reduces to

$$12x^2 - 10x'y' - 2y'^2 = 0.$$

Example 58:

Find the equations of the straight lines bisecting the angles between the pairs of straight lines given in,

(i) $4x^2 - 24xy + 11y^2 = 0$

(ii) $33x^2 - 71xy - 14y^2 = 0$

(iii) $x^2 - 2xy \sec\theta + y^2 = 0$

(iv) $x^2 + 2xy \cot\theta + y^2 = 0$

Solution:

(i) Equation of the bisectors of the angles between

$4x^2 - 24xy + 11y^2 = 0$ will be

$$\frac{x^2 - y^2}{4 - 11} = \frac{xy}{-12};$$

$$12x^2 - 7xy - 12y^2 = 0.$$

(ii) Equation of the bisectors of the angles between

$33x^2 - 71xy - 14y^2 = 0$ is

$$\frac{x^2 - y^2}{33 + 14} = \frac{xy}{-\frac{71}{2}}$$

or $71x^2 + 94xy - 71y^2 = 0.$

(iii) Equations of the lines bisecting the angles between $x^2 + 2xy \sec\theta + y^2 = 0$ will be $\dfrac{x^2 - y^2}{1 - 1} = \dfrac{xy}{\sec\theta}$ As R.H.S. is finite, denominator on L.H.S. is zero; hence the numerator of L.H.S. must be zero. Therefore the required equation is $x^2 - y^2 = 0$.

(iv) The equation of the bisectors of the angles between $x^2 + 2xy \cot\theta + y^2 = 0$ is $\dfrac{x^2 - y^2}{1 - 1} = \dfrac{xy}{\cot\theta}$

Hence as in part (3), the required equation is $x^2 - y^2 = 0$.

Example 59:

Find what straight lines are represented by the following equation and determine the angles between them

$$x^2 + 2xy \sec\theta + y^2 = 0.$$

Solution:

$x^2 + 2xy \sec\theta + y^2 = 0.$

$\Rightarrow m^2 + 2m \sec\theta + 1 = 0$ (dividing by x^2 and putting $y/x = m$)

$$\Rightarrow \quad m = \frac{-2\sec\theta \pm \sqrt{(4\sec^2\theta - 4)}}{2} = -\sec\theta \pm \tan\theta = \frac{-1 \pm \sin\theta}{\cos\theta}$$

or $\dfrac{y}{x} = \dfrac{-1 \pm \sin\theta}{\cos\theta}$

Hence the equations are $y\cos\theta + x(1 + \sin\theta) = 0$

and $y\cos\theta + x(1 - \sin\theta) = 0$.

If ϕ be the angle between them, $\tan\phi = \dfrac{2\sqrt{(\sec^2\theta - 1)}}{1+1} = \tan\theta.$

$$\phi = \theta.$$

Example 60:

Show that the equation of the straight line passing through the point (a $\cos^3\theta$, a $\sin^3\theta$) and perpendicular to the line

$\alpha \sec\theta + y \operatorname{cosec}\theta = a$ *is*

$x\cos\theta - y\sin\theta = \cos 2\theta.$

Solution:

The equation of the straight line perpendicular to the given straight line is given as

$$x \operatorname{cosec}\theta - y\sec\theta + \lambda = 0 \qquad \text{...(i)}$$

It passes through the point (a $\cos^3\theta$, a $\sin^3\theta$). Then we have

$$a\cos^3\theta - \operatorname{cosec}\theta - a\sin^3\theta\sec\theta + \lambda = 0$$

$$\Rightarrow \quad \lambda = a\left(\frac{\sin^3\theta}{\cos\theta} - \frac{\cos^3\theta}{\sin\theta}\right) = \frac{-a\cos 2\theta}{\cos\theta\sin\theta}$$

Substituting λ in (i) we have

$$x \operatorname{cosec}\theta - y\sec\theta = \frac{a\cos 2\theta}{\cos\theta\sin\theta}$$

$$x\cos\theta - y\sin\theta\, a\cos 2\theta = 0$$

Example 61:

For what value of x, the line joining (–5, 7) and (0, – 2) is perpendicular to the line joining (1, – 3) and (4, x).

Solution:

Equation of the line joining (–5, 7) and (0, –2) is given as

$$y - 7 = \frac{7+2}{-5-0}(x+5)$$

$$\Rightarrow \quad y - 7 = \left(\frac{9}{-5}\right)(x+5). \qquad \text{...(i)}$$

Equation of the line joining (1, –3) and (4, x) is

$$y + 3 = \frac{-3-x}{1-4}(x-1). \qquad ...(ii)$$

The lines will be perpendicular if the product of the slopes is (–1). This gives

$$\left(\frac{9}{-5}\right)\left(\frac{3+x}{3}\right) = -1$$

$$\Rightarrow \qquad x = -4/3. \qquad \textbf{Ans.}$$

Example 62:

Find what straight lines are represented by the following equations and determine the angles between them

$$x^2 + 2xy \cot\theta + 1 = 0.$$

Solution:

(dividing by x^2 and putting $y/x = m$)

$$\Rightarrow \quad m = \frac{-2\cot\theta \pm \sqrt{(4\cot^2\theta - 4)}}{2} = \frac{-2\cot\theta \pm 2\sqrt{\left(\frac{\cos^2\theta}{\sin^2\theta} - 1\right)}}{2}$$

$$\Rightarrow \quad \frac{y}{x} = -\frac{\cos\theta}{\sin\theta} \pm \sqrt{\left(\frac{\cos^2\theta - \sin^2\theta}{\sin^2\theta}\right)} = -\frac{\cos\theta}{\sin\theta} \pm \frac{\sqrt{(\cos 2\theta)}}{\sin\theta}$$

$$= -\frac{\cos\theta \pm \sqrt{(\cos 2\theta)}}{\sin\theta}$$

Hence the equations are $y \sin\theta + x\cos\theta - x\sqrt{(\cos 2\theta)} = 0$

and $\qquad y\sin\theta + x\cos\theta + x\sqrt{(\cos 2\theta)} = 0.$

If ϕ be the angle between the lines, then

$$\tan\theta = \frac{2\sqrt{(\cot^2\theta - 1)}}{1+1} = \sqrt{\left(\frac{\cos^2\theta - \sin^2\theta}{\sin^2\theta}\right)} = \frac{\sqrt{(\cos 2\theta}}{\sin\theta}$$

$$= \text{cosec}\,\theta\ \sqrt{(\cos 2\theta)}.$$

$$\therefore \qquad \phi = \tan^{-1}[\text{cosec}\,\theta\ \sqrt{(\cos 2\theta)}].$$

$$= 2\tan\alpha\sqrt{[(\sin^2\theta + \cos^2\theta\sin^2\alpha)\{\sin^2\theta + \cos^2\theta(1-\cos^2\alpha)]}$$

$$= 2\tan\alpha\sqrt{\{(\sin^2\theta + \cos^2\theta\sin^2\alpha)\{\sin^2\theta + \cos^2\theta.\sin^2\alpha)\}}$$

$= \tan\alpha\,(\sin^2\theta + \cos^2\theta\,\sin^2\alpha)$

Denominator $= 2(\cos^2\theta\,\sin^2\theta + \sin^2\theta - \tan^2\alpha)$

$$= \frac{1}{\cos^2\alpha}\left[2\cos^2\theta\sin^2\alpha\,\cos^2\alpha + \sin^2\theta\cos^2\alpha - \sin^2\alpha\right]$$

$$= \frac{1}{\cos^2\alpha}\left[2\cos^2\theta\sin^2\alpha\,\cos^2\alpha + \sin^2\theta\cos^2\alpha - \cos^2\theta\sin^2\alpha - \sin^2\theta\,\sin^2\alpha\right]$$

Rearranging and factorizing, we get

$$= \frac{1}{\cos^2\alpha}\left[\sin^2\theta\left(\cos^2\alpha - \sin^2\alpha\right) + \cos^2\theta\,\sin^2\alpha\left(2\cos^2\alpha - 1\right)\right]$$

$$= \frac{\cos^2\alpha - \sin^2\alpha}{\cos^2\alpha}\left[\sin^2\theta + \cos^2\theta\,\sin^2\alpha\right]$$

$$(\because 2\cos^2\alpha - 1 = \cos^2\alpha - \sin^2\alpha)$$

$= (1 - \tan^2\alpha)\,[\sin^2\theta + \cos^2\theta\,\sin^2\alpha]$.

Hence

$$\frac{\text{Numerator}}{\text{Denominator}} = \frac{2\tan\alpha\left(\sin^2\theta + \cos^2\theta\,\sin^2\alpha\right)}{\left(1 - \tan^2\alpha\right)\left(\sin^2\theta + \cos^2\theta\sin^2\alpha\right)} = \tan 2\alpha.$$

Therefore $\tan\phi = \tan 2\alpha$ by (1) or $\phi = 2\alpha$. **Proved.**

Example 4(a):

Prove that the two straight lines $(x^2 + y^2)(\cos^2\theta\sin^2\alpha + \sin^2\theta) = (x\tan\alpha - y\sin\theta)^2$ include an angle 2α.

Solution:

The equation of the lines is given as

$$(x^2 + y^2)(\cos^2\theta\,\sin^2\theta + \sin^2\theta = (x\tan\alpha - y\sin\theta)^2$$

$$\Rightarrow x^2(\cos^2\theta\,\sin^2\alpha + \sin^2\theta - \tan^2\alpha) + y^2(\cos^2\theta\,\sin^2\alpha + \sin^2\theta - \sin^2\theta) + 2xy\tan\alpha\sin\theta = 0.$$

$$\Rightarrow x^2(\cos^2\theta\,\sin^2\alpha + \sin^2\theta - \tan^2\alpha) + y^2\cos^2\theta\,\sin^2\alpha + 2x\tan\alpha\sin\theta = 0.$$

If ϕ be the angle between the lines represented by the equation.

$$\tan\phi = \frac{2\sqrt{\left[\left\{\tan^2\alpha\,\sin^2\theta - \left(\cos^2\theta\,\sin^2\alpha + \sin^2\theta - \tan^2\alpha\right)\cos^2\theta\,\sin^2\alpha\right\}\right]}}{\cos^2\theta\,\sin^2\alpha + \sin^2\theta - \tan^2\alpha + \cos^2\theta\,\sin^2\alpha} \quad ...(1)$$

Taking R.H.S.

Numerator

$$= 2\sqrt{\{\tan^2 \alpha.\sin^2 \theta - (\cos^2 \theta \sin^2 \alpha + \sin^2 \theta - \tan^2 \alpha).\cos^2 \theta \sin^2 \alpha\}}$$

$$2\sqrt{\{\tan^2 \alpha (\sin^2 \theta - \cos^4 \theta \sin^2 \theta \cos^2 \alpha - \sin^2 \theta \cos^2 \theta \cos^2 \alpha + \cos^2 \theta \sin^2}$$

[∴ $\cos^2\theta \sin^2\alpha = \cos^2 \theta \tan^2 \alpha. \chi o\sigma^2\alpha$]

Rearranging and factorizing, we get

$$= 2\tan = 2\tan\alpha\sqrt{\{(\sin^2 \theta + \cos^2 \sin^2 \alpha) (1 - \cos^2 \theta \cos^2 \alpha)]}$$

$$= 2\tan\alpha\sqrt{\{(\sin^2 \theta + \cos^2 \sin^2 \alpha)(\sin^2 \theta + \cos^2 \theta - \cos^2 \theta \cos^2 \alpha)\}}$$

[∴ $1 = \cos^2\theta + \sin^2 \theta$]

Example 63:

Find what straight lines are represented by the following equation and determine the angle between them

$$x^2 - 7xy + 12y^2 = 0.$$

Solution:

The given equation is $x^2 - 7xy + 12y^2 = 0$. ... (1)

Factorizing, we get $(x - 3y)(x - 4y) = 0$.

Hence the equations represented by (1) are

$$x - 3y = 0 \text{ and } x - 4y = 0.$$

If θ is the angle between the line then

$$\tan \theta = \frac{2\sqrt{\left(\frac{7}{2}\right)^2 - 1.12}}{1+12} = \frac{1}{13} \text{ (taking + ve sign)}$$

$$\Rightarrow \theta = \tan^{-1}\left(\frac{1}{13}\right).$$

Example 64:

Show that the two straight lines $x^2 (\tan^2 \theta + \cos^2 \theta) - 2xy \tan \theta + y^2 \sin^2 \theta = 0$, make with the axis of x angles such that the difference of their tangents is 2.

Solution:

The given equation is

$$x^2 (\tan^2\theta + \cos^2\theta) - 2xy \tan \theta + y^2 \sin^2 \theta = 0. \quad ... (1)$$

Let the equation (1) represent two straight lines $y = m_1x$ and $y = m_2x$ where m_1 and m_2 are their slopes since (1) represents two straight lines passing through origin.

The combined equation will be

$$(y - m_1x)(y - m_2x) = 0.$$

$$\Rightarrow \quad y^2 - (m_1 + m_2)\,xy + m_1m_2x^2 = 0 \quad \ldots (2)$$

As (1) (2) represent same straight lines, the coefficients of x^2, y^2 and xy must be proportional

$$\text{i.e.} \quad \frac{1}{\sin^2\theta} = \frac{-(m_1 + m_2)}{-2\tan\theta} = \frac{\tan^2\theta + \cos^2\theta}{m_1 m_2}$$

$$\text{Hence } m_1 + m_2 = \frac{2\tan\theta}{\sin^2\theta} = \frac{2}{\sin\theta\cos\theta}$$

Therefore $(m_1 - m_2)^2 = (m_1 + m_2)^2 - 4m_1m_2$

$$= \frac{4}{\sin^2\theta\cos^2\theta} - \frac{4(\sin^2\theta + \cos^4\theta)}{\sin^2\theta\cos^2\theta} = 4\left[\frac{1 - \sin^2\theta - \cos^4\theta}{\sin^2\theta\cos^2\theta}\right]$$

$$= 4\left[\frac{\cos^2\theta - \cos^4\theta}{\sin^2\theta\cos^2\theta}\right] = \frac{4\cos^2\theta(1 - \cos^2\theta)}{\cos^2\theta\sin^2\theta} = \frac{4\cos^2\theta\sin^2\theta}{\cos^2\theta\sin^2\theta}$$

Hence $\quad (m_1 - m_2)^2 = 4$

or $\quad m_1 - m_2 = \pm 2$

or $\quad \tan\theta_1 \sim \tan\theta_2 = 2,$

where θ_1 and θ_2 are the inclinations of the lines represented by (1) from the x-axis.

Example 65:

Prove that the two straight lines

$x^2 \sin^2\alpha \cos^2\theta + 4xy \sin\alpha \sin\theta + y^2 [4\cos\alpha - (1 + \cos\alpha)^2 \cos^2\theta] = 0$ meet at an angle α.

Solution:

The lines are given by

$$x^2 \sin^2\alpha \cos^2\theta + 4xy \sin\alpha \sin\theta + y^2$$
$$[4\cos\alpha - (1 + \cos\alpha)^2 \cos^2\theta] = 0.$$

If ϕ be the angle between them, then we have

$$\tan\phi = \left\{\frac{2\sqrt{[(2\sin\alpha\sin\theta)^2 - \sin^2\alpha\cos^2\theta\,[4\cos\alpha - (1+\cos\alpha)^2\cos^2\theta)]}}{\sin^2\alpha\cos^2\theta + \left[4\cos\alpha - (1+\cos\alpha)^2\cos^2\theta\right]}\right\} \quad \dots (1)$$

Taking R.H.S.

Numerator

$= 2\sqrt{[4\sin^2\alpha\sin^2\theta - (\sin^2\alpha\cos^2\theta)\{4\cos\alpha - (1+\cos\alpha)^2\cos^2\theta\}]}$

$= 2\sqrt{[\sin^2\alpha\{4\sin^2\theta - 4\cos\alpha\cos^2\theta + (1+\cos\alpha)^2\cos^4\theta\}]}$

$= 2\sqrt{[\sin^2\alpha\{4 - 4\cos^2\theta - 4\cos\alpha\cos^2\alpha\cos^2\theta + (1+\cos\alpha)^2\cos^4\theta\}]}$

$[\therefore \sin^2\theta = 1 - \cos^2\theta]$

$= 2\sqrt{[\sin^2\alpha\{4 - 4\cos^2\theta(1+\cos\alpha) + (1+\cos\alpha)^2\cos^4\theta)]}$

$= 2\sqrt{[\sin^2\alpha\{2 - (1+\cos\alpha)\cos^2\theta\}^2]}$

$= 2\sin\alpha\,[2 - (1+\cos\alpha)\cos^2\theta].$

Denominator $= \sin^2\alpha\cos^2\theta + 4\cos\alpha - (1+\cos\alpha)^2\cos^2\theta.$

Rearranging and simplifying, we have

$= 4\cos\alpha - (1 + \cos^2\alpha + 2\cos\alpha)\cos^2\theta + \sin^2\alpha\cos^2\theta$

$= 4\cos\alpha - \cos^2\theta\,[1 + \cos^2\alpha + 2\cos\alpha - \sin^2\theta].$

$= 4\cos\alpha - \cos^2\theta.[2\cos^2\alpha + 2\cos\alpha] \quad (\therefore 1 - \sin^2\alpha = \cos^2\alpha)$

$= 4\cos\alpha - \cos^2\theta.\ 2\cos\alpha(\cos\alpha + 1)$

$= 2\cos\alpha\,[2 - (1+\cos\alpha).\cos^2\theta].$

$$\therefore \quad \frac{\text{Numerator}}{\text{Deno min ator}} = \frac{2\sin\alpha\left[2-(1+\cos\alpha)\cos^2\theta\right]}{2\cos\alpha\left[2-(1+\cos\alpha)\cos^2\theta\right]} = \tan\alpha.$$

Hence (1), $\tan\phi = \tan\alpha$ or $\phi = \alpha$.

Example 66:

Prove that the following equation represents two straight lines; find also their point of intersection and the angle between them

$$6y^2 - xy - x^2 + 30y + 36 = 0.$$

Solution:

Here we have

$$6y^2 - xy - x^2 + 30y + 36 = 0.$$

Hence a = –1, b = 6, c = 36, h = –1/2, g = 0, f = 15,

$\Delta = abc + 2fgh - af^2 - bg^2 - ch^2$

$$= (-1).6.36 + 2.15.0\left(-\frac{1}{2}\right) - (-1)\,(15)^2 - 6.(0)^2 - 36\left(-\frac{1}{2}\right)^2 = 0$$

Hence the equation represents two straight lines.

Again $6y^2 - xy - x^2 = (3y + x)\,(2y - x)$

Hence let

$6y^2 - xy - x^2 + 30y + 36 \equiv (3y + x + A) \times (2y - x + B)$.
...(1)

Comparing coefficients of x and y, we get $B - A = 0$ ...(2)

$3B + 2A = 30$ ...(3)

and $2y - x + 6 = 0$. ...(5)

If ϕ be the angle between them

$$\tan\theta = \frac{2\sqrt{\left[\left(-\frac{1}{2}\right)^2 - (-1)\,(6)\right]}}{-1+6} = \frac{2\sqrt{\left(\frac{25}{4}\right)}}{5} = 1.$$

$\therefore \quad \phi = 45^\circ$.

Solving (4) and (5), we get $x = \frac{6}{5}, y = -\frac{12}{5}$.

So point of intersection is $\left(\frac{6}{5}, -\frac{12}{5}\right)$.

Example 67:

Prove that the following equation represents two straight lines; find also their point of intersection and the angle between them

$$x^2 - 5xy + 4y^2 + x + 2y - 2 = 0.$$

Solution:

Here we have

$$x^2 - 5xy + 4y^2 + x + 2y - 2 = 0.$$

No we here $a = 1, b = 4, c = -2, h = -\frac{5}{2}, g = \frac{1}{2}, f = 1$

$$\therefore \Delta = 1.4\,(-2) + 2.1.\frac{1}{2}.-\left(\frac{5}{2}\right) - 1.(1)^2 - 4\left(\frac{1}{2}\right)^2 - (-2)\left(-\frac{5}{2}\right)^2 = 0.$$

Hence it represents two straight lines.

Again now $x^2 - 5xy + 4y^2 = (x - 4y)(x - y)$.

So let $x^2 - 5xy + 4y^2 + x + 2y - 2 \equiv (x - 4y + A)(x - y + B)$.

Comparing the coefficients of x and y, we get

$$A + B = 1$$

and $-A-4B = 2$.

Solving, we get $A = 2$, $B = -1$.

Putting in (1), the required equations are

$$x - 4y + 2 = 0 \quad \ldots(2)$$

and $x - y - 1 = 0$. ... (3)

Solving (2) and (3), the point of intersection is (2, 1).

If θ is the angle between them, then we have

$$\tan\theta = \frac{2\left\{\sqrt{\left(-\frac{5}{2}\right)^2 - 1.4}\right\}}{1+4} = \frac{3}{5} \therefore \theta = \tan^{-1}\left(\frac{3}{5}\right).$$

Example 68:

Find the value of k so that the following equation may represent pairs of straight lines $12x^2 + xy - 6y^2 - 29x + 8y + k = 0$.

Solution:

The given equations $12x^2 + xy - 6y^2 - 29x + 8y + k = 0$

Here $a = 12$, $b = -6$, $c = k$, $f = 4$, $g = -\frac{29}{2}$, $h = \frac{1}{2}$

As it represents two straight lines, Δ must be zero.
Applying the condition $\Delta = 0$ we have

$$\Delta = 12(-6)k + 24\left(-\frac{29}{2}\right)^2\left(\frac{1}{2}\right) - 12(4)^2 - (-6)\left(-\frac{29}{2}\right)^2 - k\left(\frac{1}{2}\right)^2 = 0$$

$$\Rightarrow -72k - 58 - 192 + \frac{2523}{2} - \frac{k}{4} = 0.$$

Solving $\quad k = 14$.

Example 69:

Find the value of k so that the following equation may represent pairs of straight lines

$$2x^2 + xy - y^2 + kx + 6y - 9 = 0.$$

Solution:

The given equation is

$$2x^2 + xy - y^2 + kx + 6y - 9 = 0$$

$$\Rightarrow \quad 4x^2 + 2xy - 2y^2 + 2kx + 12y - 18 = 0.$$

Here a = 4, b = –2, c = –18, f = 6, g = k, h = 1.

As it represents two straight lines, Δ must be zero.

$$\therefore \Delta = 4(-2)(-18) + 2(6)k.1 - 4(6)^2 (-2)(k)^2 - (18)(1)^2 = 0$$

$$\Rightarrow \quad 144 + 12k - 144 + 2k^2 + 18 = 0$$

$$\Rightarrow \quad k^2 + 6k + 8 = 0 \text{ or } (k + 3)^2 = 0.$$

Hence $k = -3$. **Ans.**

Example 70:

Prove that the equation $x^2 + 6xy + 9y^2 + 4x + 12y - 5 = 0$ represents two parallel lines.

Solution:

The given equation is

$$x^2 + 6xy + 9y^2 + 4x + 12y - 5 = 0. \quad ...(1)$$

Here a = 1, b = 9, c = –5, f = 6, g = 2, h = 3. Hence

$$\Delta = 1.9(-5) + 2.6.2.3. -1.(6)^2 -9(2)^2 - (-5)(3)^2 = 0.$$

So the equation represents 2 straight lines.

Again $h^2 - ab = (3)^2 - 1.9 = 0$; so $h^2 = ab$.

Therefore, the lines represents by (1) are parallel.

Example 71:

Find the value of k so that following equation may represent pairs of straight lines $6x^2 + 11xy - 10y^2 + x + 31y + k = 0$.

Solution:

The given equation is

$$6x^2 + 11xy - 10y^2 + x + 31y + k = 0.$$

It will represent two straight lines if Δ = 0.

Here a = 6, b = –10, c = k, $f = \frac{31}{2}$, $f = \frac{1}{2}$, $h = \frac{11}{2}$

Now, $\Delta = 6\,(-10).k + 2.\frac{31}{2}, \frac{1}{2}, \frac{11}{2} - 6\left(\frac{31}{2}\right)^2 - (-10)\left(\frac{1}{2}\right)^2 - k.\left(\frac{11}{2}\right)^2 = 0.$

$$\Rightarrow \quad -60k + \frac{341}{2} - \frac{2883}{2} + \frac{5}{2} - \frac{121}{4}k = 0.$$

Solving, k = –15. **Ans.**

Example 72:

Prove that the following equation represents two straight lines; find also their point of intersection and the angle between them

$$3y^2 - 8xy - 3x^2 - 29x^2 - 29x + 3y - 18 = 0.$$

Solution:

The given equation is

$$3y^2 - 8xy - 3x^2 - 29x + 3y - 18 = 0.$$

Here a = –3, b = 3, c = –18, $f = \frac{3}{2}, g = \frac{-29}{\cdot 2}, h = -4$

Hence we have

$$\Delta = (-3)\,(3)\,(-18) + 2\left(\frac{3}{2}\right)\left(-\frac{29}{2}\right)(-4) - (5)\left(\frac{3}{2}\right)^2 - (3)\left(-\frac{29}{2}\right)^2 - (-18)\,(-4)^2 - = 0.$$

Hence two straight lines are represented by

Now $3y^2 - 8xy - 3x^2 = (3y + x)(y - 3x)$. Hence let

$3y^2 - 8xy - 3x^2 - 29x + 3y - 18 \equiv (3y + x + A)(y - 3x + B)$ ...(1)

Equating the coefficients of x and y, we get

$$-3A + B = -29$$

and A + 3B = 3.

Solving, we get A = 9 and B = –2.

Substituting in (1), and equating each to zero, the equations are

$$3y + x + 9 = 0 \quad ...(2)$$

and y – 3x – 2 = 0. ...(3)

Solving (2) and (3), we get the point of intersection as $\left(-\frac{3}{2}, -\frac{5}{2}\right)$

If θ is the angle between them, then we have

$$\tan\theta = \frac{1\sqrt{\{(-4)^2 - (-3)\,(3)\}}}{-3+3} = \infty;\ \theta = 90^\circ$$

Example 73:

If the pairs of straight lines $x^2 - 2pxy - y^2$ 0 and $x^2 - 2qxy - y^2$ 0 be such that each pair bisects the angle between the other pair, prove that $pq = -1$

Solution:

The given pairs of straight lines are

$$x^2 - 2pxy - y^2 = 0 \quad \text{...(i)}$$

$$x^2 - 2qxy - y^2 = 0 \quad \text{...(ii)}$$

Equation of the bisectors of angles of (i) are given by

$$\frac{x^2 - y^2}{1-(-1)} = \frac{xy}{-p} \left(\text{using the equation } \frac{x^2 - y^2}{a - b} = \frac{xy}{h} \right)$$

$$\Rightarrow \quad x^2 + \left(\frac{2}{p}\right) xy - y^2 = 0. \quad \text{...(iii)}$$

It is given that (ii) represents the equation of the bisectors of angles of (i). ∴ equation (ii) must represents the same pair of bisectors. This gives on comparison

$$\frac{1}{1} = \left(\frac{2/p}{-2q}\right) = \frac{1}{1} \text{ or } \frac{1}{p} = -q \text{ or } pq = -1,$$

which is the required result.

Example 74:

Prove that the general equation $ax^2 + 2hxy + by^2 + 2gx + 2fy + c = 0$ represents two parallel straight lines if $h^2 = ab$ and $bg^2 = af^2$. Also show that the distance between them is

$$2\sqrt{\frac{g^2 - ac}{a(a+b)}}.$$

Solution:

The given equation is

$$ax^2 + 2hxy + by^2 + 2gx + 2fy + c = 0. \quad \text{...(i)}$$

Multiplying (i) by a we get

$$a^2x^2 + 2ax(hy + g) = -aby^2 - 2afy - ca$$

or $(ax + hy + g)^2 = (hy + g)^2 - aby^2 - 2afy - ca$

$$= (h^2 - ab)\, y^2 + 2(hg - af)\, y + g^2 - ca$$

$$(ax + hy + g) = \pm \sqrt{\left\{\left(h^2 - ab\right)y^2 + 2\left(hg - af\right)y + \left(g^2 - ca\right)\right\}}$$

...(ii)

If (i) represents two parallel straight lines, then the right hand side of (ii) must not contain any term involving y.

This gives

$$h^2 - ab = 0 \qquad \text{...(iii)}$$

and $hg - af = 0$ ...(iv)

(iv) can also be expressed as

$$h^2g^2 = a^2f^2$$

$$\Rightarrow \quad abg^2 = a^2f^2$$

$$\Rightarrow \quad bg^2 = af^2. \qquad \text{...(v)}$$

Under these conditions equation (ii) reduces to

$$ax + hy + g = \pm \sqrt{\left(g^2 - ca\right)}$$

which gives two parallel lines as

$$ax + hy + g + \sqrt{g^2 - ca} = 0. \qquad \text{...(vi)}$$

and $ax + hy + g - \sqrt{g^2 - ca} = 0.$ ...(vii)

Distance between these parallel lines is equal to the perpendicular distance from any point on one line to second line. Any point on line (vi) is $\left\{0, -\dfrac{(g + \sqrt{g^2 - ca})}{h}\right\}$

The perpendicular distance from this point on lines (vii) is

$$\frac{\left\{\dfrac{-(g + \sqrt{g^2 - ca})}{h}\right\} + g - \sqrt{\left(g^2 - ca\right)}}{\sqrt{\left(a^2 + h^2\right)}}$$

$$\Rightarrow \quad \frac{-2\sqrt{g^2 - ca}}{\sqrt{\left(a^2 + h^2\right)}} = \frac{-2\sqrt{g^2 - c^2}}{\sqrt{\left(a^2 + ab\right)}}$$

Hence the require distance between these parallel line is

$$\frac{\sqrt{g^2 - ca}}{\sqrt{a\left(a+b\right)}}.$$

Example 75:

If the lines $ax^2 + 2hxy + by^2 = 0$ *be the sides of a parallelogram and the line* $lx + my = 1$ *be one of its diagonals, show that the equation of the other diagonal is* $y\ (bl - hm) = x\ (am - hl)$

Solution:

Since the other diagonal will pass through the origin, let its equation be

$$y = m'x. \qquad \text{...(i)}$$

The point of intersection of these two diagonals is

$$\left(\frac{1}{l + mm'}, \frac{m'}{l + mm'}\right).$$

The point of intersection of the line $lx + my = 1$ and $ax^2 + 2hxy + by^2 = 0$ are given by

$$ax^2 + 2hx\left(\frac{1 - lx}{m}\right) + b\left(\frac{1 - lx}{m}\right)^2 = 0$$

$$\Rightarrow \quad (am^2 - 2hm + bl^2)\ x^2 + 2(hm - bl)\ x + b = 0.$$

This gives

$$x_2 + x_2 = -\frac{2\left(hm - bl\right)}{am^2 - 2hlm + bl^2}, \qquad \text{...(ii)}$$

where x_1 and x_2 are the abscissas of the points of intersection of the curve $ax^2 + 2hxy + by^2$ and $lx + my = 1$.

Since the diagonals of parallelogram bisect each other, it gives

$$\frac{1}{l + mm'} = \frac{bl - hm}{am^2 - 2hlm + bl^2}$$

$$\therefore \quad m' = \frac{am - hl}{bl - hm}.$$

Substituting m' in (1), the equation of other diagonal becomes

$$y = \frac{am - hl}{bl - hm}x$$

or $y(bl - hm) = x(am - hl)$,

which is the required result.

Example 76:

Prove that the following equation represents two straight lines; find also their point of intersection and the angle between them

$$y^2 + xy - 2x^2 - 5x - y - 2 = 0.$$

Solution:

The given equation is

$$y^2 + xy - 2x^2 - 5x - y - 2 = 0.$$

Here $a = -2$, $b = 1$, $c = -2$, $f = -\frac{1}{2}$, $g = -\frac{5}{2}$, $h = \frac{1}{2}$.

Hence

$$\Delta = (-2)\ (1)\ (-2) + 2.$$

$$\left(-\frac{12}{2}\right)\left(-\frac{5}{2}\right)\left(\frac{1}{2}\right) - (-2)\left(-\frac{1}{2}\right)^2 - 1.\left(-\frac{5}{2}\right)^2 - (-2).\left(\frac{1}{2}\right)^2 = 0.$$

So the equation represents two straight lines are given as

Now, $y^2 + xy - 2x^2 = (y - x)\ (y + 2x)$. Hence let

$$y^2 + xy - 2x^2 - 5x - y - 2 \equiv (y - x + A)\ (y + 2x + B). \quad \ldots(1)$$

Comparing the coefficients of y and x, we get

$$A + B = -1 \text{ and } 2A - B = -5.$$

Solving, we get $A = -2$, $B = 1$.

Substituting in (1) and putting each factor equal to zero, the equations are

$$y - x - 2 = 0 \quad \ldots(2)$$

and $y + 2x + 1 = 0$. ....(3)

Solving (2) and (3), the point of intersection is $(-1, 1)$.

If θ be the angle between them, then we have

$$\tan\theta = \frac{2\sqrt{\{(1/2)^2 - (-2)(1)\}}}{-2-1} = 3 \text{ (omitting --ve sign)}$$

or $$\theta = \tan^{-1} 3.$$

Important Condition. If the two lines represented by $ax^2 + 2hxy + by^2 + 2gx + 2fy + c = 0$, be perpendicular to each other, then $a + b = 0$ and if, they are parallel, then $h^2 = ab$ and conversely, if $a + b = 0$, the lines are perpendicular and if $h^2 = ab$, the lines are parallel.

Example 77:

Find the value of k so that the following equation may present pairs of straight lines $12x^2 - 10xy + 2y^2 + 11x - 5y + k = 0$.

Solution:

The given equation is

$$12x^2 - 10xy + 2y^2 + 11x - 5y + k = 0.$$

Here $a = 12$, $b = 2$, $c = k$, $f = -\frac{5}{2}$, $g = \frac{11}{2}$, $h = -5$.

If it represents 2 straight lines, $\Delta = 0$.

$$\Delta = 12.2k + 2.\left(-\frac{5}{2}\right)\left(\frac{11}{2}\right)(-5) - 12\left(-\frac{5}{2}\right)^2 - 2\left(\frac{11}{2}\right)^2 - k(-5)^2 = 0.$$

$\Rightarrow 24k + \frac{275}{2} - 75 - \frac{121}{2} - 25k = 0.$ Solving $k = 2$. **Ans.**

Example 78:

Find the value of k so that the following equation may represent pairs of straight lines $12x^2 + kxy + 2y^2 + 11x - 5y + 2 = 0$.

Solution:

The given equation is

$$12x^2 + kxy + 2y^2 + 11x - 5y + 2 = 0.$$

$$\Rightarrow \quad 24x^2 + 2kxy + 4y^2 + 22x - 10y + 4 = 0.$$

Now $a = 24$, $b = 4$, $c = 4$, $f = -5$, $g = 11$, $h = k$.

If it represents 2 straight lines, $\Delta = 0$.

$$\Delta = 24.4.4 + 2.(-5)\,11.k - 24.(-5)^2 - 4\,(11)^2 - 4k^2 = 0.$$

$$\Rightarrow \quad -4k^2 - 110k - 700 = 0$$

$$\Rightarrow \quad 2(k + 10)(2k + 35) = 0$$

Hence either k + 10 or 2x + 35 = 0

$$\therefore\ k = -10 \text{ or } k = -\frac{35}{2}.$$

Example 79:

Find the value of k so that the following equation may represent pairs of straight lines $x^2 + kxy + y^2 - 5x - 7y + 6 = 0$.

Solution:

The given equation is

$$x^2 + kxy + y^2 - 5x - 7y + 6 = 0.$$

$$\Rightarrow\ 2x^2 + 2kxy + 2y^2 - 10x - 14y + 12 = 0.$$

Here a = 2, b = 2, c = 12, f = –7, g –5, h = k.

If it represents 2 straight lines, $\Delta = 0$.

$$\therefore\ \Delta = 2.2.12 + 2(-7)(-5)k - 2(-7)^2 - (-5)^2 - 12k^2 = 0$$

$$\Rightarrow\ -12k^2 + 70k - 100 = 0$$

or $\quad 6k^2 - 35k + 50 = 0$

$$\Rightarrow\ (3k - 10)(2k - 5) = 0$$

or $\quad k = 10/3$ or $5/2$.

$k = 5/2$ **Ans.**

Example 80:

Prove that the equations to the straight lines passing through the origin which make an angle α with the straight line $y + x = 0$ are given by the equation $x^2 + 2xy \sec 2\alpha + y^2 = 0$.

Solution:

Equation of any line passing through origin is y = mx. ...(1)

The given equation is y + x = 0 ...(2)

If α is the angle between (1) and (2), then $\tan\alpha = \dfrac{-1-m}{-1+m}$

[$\therefore$ slope of (2) is –1]

$$\Rightarrow\ m = \frac{\tan\alpha - 1}{\tan\alpha + 1} = -m_1 \text{ (say)}. \quad ...(3)$$

Again, if the angle between (1) and (2) be $-\alpha$, then $-\tan\alpha = \dfrac{-1-m}{-1+m}$

$$\text{or}\quad m = \frac{\tan\alpha + 1}{\tan\alpha - 1} = m_2 \text{ (say)}. \quad ...(4)$$

Putting the values of m_1 and m_2 from (3) and (4) in (1), the two equations are

$$\left(y - \frac{\tan\alpha - 1}{\tan\alpha + 1}\right) = 0 \text{ and } \left(y - \frac{\tan\alpha + 1}{\tan\alpha - 1}x\right) = 0$$

So combined equation is $\left(y - \dfrac{\tan\alpha - 1}{\tan\alpha + 1}x\right)\left(y - \dfrac{\tan\alpha + 1}{\tan\alpha - 1}x\right) = 0$

$\Rightarrow \; y^2 + 2xy \sec 2\alpha + x^2 = 0$

$$\Rightarrow \quad \left[\therefore -\left(\frac{\tan\alpha - 1}{\tan\alpha + 1} + \frac{\tan\alpha + 1}{\tan\alpha - 1}\right) = 2\sec 2\alpha\right].$$

Example 81:

Find the value of k so that the following equation may represent Paris of straight lines

$$6x^2 + xy + ky^2 - 11x + 43y - 35 = 0.$$

Solution:

The given equation is

$$6x^2 + xy + ky^2 - 11x + 43y - 35 = 0.$$

$\Rightarrow \; 12y^2 + 2xy + 2ky^2 - 22x + 86y = 70 = 0$ (multiplying by 2).

If it represents two straight lines, Δ must be zero.

Here $a = 12$, $b = 2k$, $c = -70$, $h = 1$. $a = 43$, $g = -11$

$\therefore \Delta = 12.2k.(-70) + 2.43, (-11)(1) - 12(43)^2 - 2k(-11)^2 - (-70)(1)^2 = 0.$

or $\; -1680k - 946 - 22188 - 242k + 70 = 0.$

Solving, we get $\quad k = -12.$ **Ans.**

Example 82:

Find the value of k so that the following equation may represent pairs of straight lines $kxy - 8x + 9y - 12 = 0$.

Solution:

The given equation is

$$kxy - 8x + 9y - 12 = 0$$

or

$$2kxy - 16x + 18y - 24 = 0.$$

Here $a = 0$, $b = 0$, $c = -24$, $f = 9$, $g = -8$, $h = k$.

As it represents 2 straight lines, $\Delta = 0$. Then we have

$\Delta = 0.0\,(-24) + 2.9.(-8)\,(k) - 0\,(9)^2 - 0(-8)^2 - (-24)\,k^2 = 0$

or $\qquad 24k^2 - 144k = 0.$

Solving, we get k = 0 or k = 6. If k = 0, it will give an equation of 1st degree.

$k = 6.$

Example 83:

Find the value of k so that the following equation may represent pairs of straight lines $x^2 + 10/3xy + y^2 - 5x - 7y + k = 0$.

Solution:

The given equation is

$$x^2 + \frac{10}{3}xy + y^2 - 5x - 7y + k = 0.$$

Here we have

$$a = 1,\ b = 1,\ c = k,\ f = -\frac{7}{2}, g = -\frac{5}{2}, h = \frac{5}{3}.$$

As it represents 2 straight lines, $\Delta = 0$.

$$\therefore\quad \Delta = 1.1.k + 2\left(-\frac{7}{2}\right)\left(-\frac{5}{2}\right)\left(\frac{5}{3}\right) - 1.\left(-\frac{7}{2}\right)^2 - 1\left(\frac{5}{2}\right)^2 - k\left(\frac{5}{3}\right)^2 = 0$$

$$\Rightarrow\quad k + \frac{175}{6} - \frac{49}{4} - \frac{25}{9}k = 0.$$

Solving $\qquad k = 6.$ **Ans.**